Hydrophobic Interactions in Food Systems

Authors

Shuryo Nakai, Ph.D.
and
Eunice Li-Chan, Ph.D.
Department of Food Science
University of British Columbia
Vancouver, B.C.
Canada

CRC Press
Taylor & Francis Group
Boca Raton London New York

CRC Press is an imprint of the
Taylor & Francis Group, an **informa** business

First published 1988 by CRC Press
Taylor & Francis Group
6000 Broken Sound Parkway NW, Suite 300
Boca Raton, FL 33487-2742

Reissued 2018 by CRC Press

Library of Congress Cataloging-in-Publication Data

Hydrophobic interactions in food systems.

Includes bibliographies and index.
1. Proteins--Surfaces. 2. Food--Protein content.
3. Surface chemistry. 4. Proteins--Structure-
activity relationships. I. Nakai, Shuryo. II. Li-Chan,
Eunice. [DNLM: 1. Dietary Proteins--metabolism.
2. Food Chemistry. 3. Structure-Activity Relationship.
QU 50 H995]
QP551.H93   1988   599'.013   87-9347
ISBN 0-8493-6044-7

A Library of Congress record exists under LC control number: 87009347

Publisher's Note
The publisher has gone to great lengths to ensure the quality of this reprint but points out that some imperfections in the original copies may be apparent.

Disclaimer
The publisher has made every effort to trace copyright holders and welcomes correspondence from those they have been unable to contact.

ISBN 13: 978-1-315-89421-8 (hbk)
ISBN 13: 978-1-351-07331-8 (ebk)

Visit the Taylor & Francis Web site at http://www.taylorandfrancis.com and the
CRC Press Web site at http://www.crcpress.com

PREFACE

The importance of hydrophobic effects in biological functions has been well recognized. However, research on the use of hydrophobic effects in quantitatively correlating with these functions is still in its infancy. Although empirical, the results of the study on quantitation so far are encouraging, and this approach has a great potential to achieve the final goal of quantitative elucidation of the structure-function relationship of food components. Charge and structural parameters, in addition to hydrophobicity, should also be taken into consideration in this attempt. However, the definition of hydrophobicity and the methodology for its quantitative evaluation are still in a state of confusion. Therefore, it is hoped that this uniscience monograph will assist scientists in the field in reviewing and up-dating their information, and will prompt them to adopt a unified quantitative approach to the study of hydrophobic interactions in food systems.

Since a CRC uniscience publication aims at presentation of information relating to important progress of timely interest in a discipline, literature cited is restricted to mostly recent publications. Despite possible criticisms for not citing many other valuable works, we decided to stay within narrow ranges in covering the literature by placing emphasis on quantitative hydrophobicity-functionality relationships, mainly to avoid the controversy, confusion, and overlap which have been plaguing the study on this specific topic.

The first part of this monograph reviews the current knowledge on (1) the definition of hydrophobic forces in comparison to other forces (2) definition of protein hydrophobicity and its analytical approach, and (3) the role of hydrophobicity in the quantitative structure-activity relationship (QSAR). The second part of the monograph presents, in some detail, an example of the application of the hydrophobicity concept to a particular food system, namely muscle proteins. The importance of hydrophobic interactions in the modification of structure-functionality relationship of food proteins is discussed in the final chapter. These last two chapters reveal that despite the overwhelming qualitative evidence of the significance of hydrophobic interactions in food protein functionality, there are currently only limited examples of the quantitative application of this knowledge.

Special credit is given with our great thanks to E. Dickson and G. Sansby, authors of *Colloids in Food* (Applied Science Publishers), for citing their explanation of the Hamaker theory, which is the most comprehensive to date.

Shuryo Nakai and Eunice Li-Chan

THE AUTHORS

Shuryo Nakai, Ph.D., is a Professor at the Department of Food Science, the University of British Columbia, Vancouver, B.C., Canada.

Dr. Nakai graduated in 1950 from the University of Tokyo with a B.Sc. in agricultural chemistry, and then worked as Division Head, at the Research Institute, of Meiji Milk Products Company, Tokyo, until he received his Ph.D. degree in dairy chemistry in 1962 from the University of Tokyo. After working for four years from 1962 as a Research Associate at the Department of Food Science, the University of Illinois, he joined the faculty at the University of British Columbia.

Dr. Nakai is a member of the Canadian Institute of Food Science and Technology, the Institute of Food Technologists, the American Chemical Society, and the American Dairy Science Association.

Dr. Nakai has published over 130 research papers, and his current research interests include structure-functionality relationships of food proteins and enzymes, computer-assisted optimization of food research and development, chemistry of immunoglobulins, pattern recognition of chromatographic data for objective flavor evaluation, and separation of bioactive compounds.

Among other awards, he has received the W. J. Eva Award of Excellence from the Canadian Institute of Food Science and Technology and the Killam Research Prize from the University of British Columbia.

Eunice Li-Chan, Ph.D., is a Research Associate at the Department of Food Science, the University of British Columbia, Vancouver, B.C., Canada.

Dr. Li-Chan received a B.Sc. in food science in 1975 from the University of British Columbia and her M.Sc. in biochemistry from the University of Alberta in 1977. She graduated from the University of British Columbia with a Ph.D. in food science in 1981 and was awarded the Killam postdoctoral fellowship from the Natural Science and Engineering Research Council of Canada for two years.

Dr. Li-Chan is a member of the Institute of Food Technologists, the Canadian Institute of Food Science and Technology, the American Chemical Society, and the New York Academy of Science.

Dr. Li-Chan has published over 25 research papers, and her current research interests include structure-function relationships of food proteins and improvement of their nutritional and functional properties by chemical, physical, or enzymatic modification.

TABLE OF CONTENTS

Chapter 1

HYDROPHOBIC INTERACTION

TABLE OF CONTENTS

I. INTRODUCTION

In this chapter, molecular forces to maintain integrity of the structure of compounds, especially proteins, are discussed first. Then, two different interpretations of the hydrophobic interaction are compared, entropic forces and solvent effects. In section IV, hydrophobicity and polarity concepts are compared and their relationship with solubility phenomenon is discussed. In the last section, the effects of salts and urea on protein hydrophobicity are explained using data from the recent literature.

II. MOLECULAR FORCES

Noncovalent forces were discovered by van der Waals in 1873 in an attempt to explain the deviation of a real gas from the ideal gas law. Noncovalent bond energies are difficult to measure quantitatively, but generally are one to three orders of magnitude smaller than covalent bonding energies.

A. Dispersion Forces

Dispersion forces occur between any pair of atoms. Each atom behaves like an oscillating dipole generated by electrons moving in relation to the nucleus of the atom. In a pair of atoms, the dipole in an atom polarizes the opposing atom. As a result, the oscillators are coupled, giving rise to an attractive force between the atoms. The theory of the attractive interaction between atoms is due to London,[1] and such forces are often called London forces, or dispersion forces. They are long-range forces, arising from attractive forces between atoms over an extended distance.

It was shown by London,[1] following the appearance of quantum theory, that this relationship is also applicable to the dispersion energy between two nonpolar molecules, due to the polarization of the electron density in one molecule by charge fluctuation in the other. Except in highly polar materials, it is the sum of London dispersion forces that contributes most to the total van der Waals attraction between macroscopic bodies.

To a good approximation, the attractive interaction energy (U) between two isolated molecules i and j is of the form:

$$U_{ij}(r) = -\Lambda_{ij}/r^6 \qquad (1)$$

where r is the distance between the centers of mass, and Λ_{ij} is a constant related to the molecular polarizabilities. This type of interaction is named after van der Waals,[2] who recognized that the effect of a dipole-dipole interaction (energy proportional to r^{-3}) when averaged over all orientations gives an energy proportional to r^{-6}.

London "dispersion" force is one in a set of van der Waals interactions. The others are the "orientations" or Keesom interaction between permanent dipoles of different molecules, and the "induction" or Debye interaction between the permanent dipole of a molecule with the dipole induced in another molecule. In general, the dispersion effect is stronger than these other two types of van der Waals interactions. The exception to this general rule is found between molecules of highly polar materials such as water. Orientation and induction account for about 80% of the van der Waals attraction between water molecules and are experimentally and theoretically reasonable to consider as the major components of hydrogen bonding. It can be concluded that orientation and induction form the major long-range components of the hydrogen bond.

B. Repulsive Forces

Short-range forces are repulsive, having their quantum mechanical origin in the overlap of electron density of adjacent molecules. Attractive and repulsive forces act simultaneously and cannot be isolated. The total interatomic potential of noncovalent interaction between nonpolar molecules is written in the form:

$$V_{LJ}(r) = A/r^{12} - B/r^6 \qquad (2)$$

which is commonly known as the Lennard-Jones (LJ) potential, or the 12-6 potential.[3] The first term in this potential represents the "repulsion" and the second the "attraction". The 12th-power dependence on distance of the first term is an empirical choice mainly for computation convenience, and devoid of theoretical significance. Various other powers have sometimes been used. The parameter B has been studied extensively in attempts to relate the strength of attraction to atomic polarizabilities. However, the complexity of the problem of interatomic attractions in polyatomic molecules makes it impractical to determine B on *a priori* theoretical grounds. Thus it is common practice to fit parameters A and B empirically, so as to obtain optimum agreement between theory and experiment.[4]

Another common representation of intermolecular or interatomic interactions is the so-called Buckingham or exp-6 potential:

$$V_B(r) = Ae^{-\alpha r} - B/r^6 \qquad (3)$$

where the repulsive term can also be written as $e^{-\alpha(r - r_0)}$. A and B are again adaptable parameters. This potential may be more accurate than the LJ potential in representing the repulsive force, but requires one more adaptable parameter α for each atom which, being in the exponent, is more difficult to fit by the least squares method.[4]

C. Electrostatic Interactions

Since covalent bonds between different types of atoms lead to an asymmetric bond electron distribution, most atoms of a molecule carry partial charges. Since a neutral molecule has no net charge, it contains only dipoles or higher multipoles. These multipoles interact with each other according to Coulomb's law:

$$E = 332/\epsilon \cdot (q_1 \cdot q_2)/r_{12} \qquad (4)$$

where q_i are the partial charges in electron charges, r_{12} is the distance between partial charges, ϵ is the dielectric constant, and E is the repulsion or attraction energy.

D. Van der Waals Potentials

It is customary to combine all three noncovalent forces, i.e., electron shell repulsion, dispersion forces, and electrostatic interactions, into a single, simple potential function (or force-field), which is called the "van der Waals potential" for historical reasons:

$$E = A/r^{12} - B/r^6 + (q_i \cdot q_j)/r \qquad (5)$$

where $q_i \cdot q_j$ is the product of the effective charge of the contact partners.

E. Salt Bridges

There are only a few salt bridges in proteins, with the exception of phosphoproteins and some glycoproteins. These salt bridges amount to about 5 kcal/mol for adjacent carboxylate and ammonium groups in medium with a dielectric constant $\epsilon = 4$.

In the solvent, charged ions attract surrounding dipoles of water molecules by an electrostatic monopole-dipole interaction. The resulting energy largely offsets the electrostatic attraction energy between the ions. Therefore, the resulting binding enthalpy is about zero.

However, the solvent entropy contribution is appreciable, favoring salt bridge formation. The resultant net free energy of transfer from a monodisperse solution in water to a solution in a nonpolar solvent is of the order of 1 kcal/mol in favor of salt bridge formation, which is relatively small.[5]

F. Hydrogen Bonds

The interatomic distances of monocovalent atomic contacts are significantly shorter than the values calculated as the sum of the corresponding van der Waals radii when one of the atoms is hydrogen. The distance between an amide H and a carbonyl 0, for example, is only 1.9 Å, shorter than 2.7 Å, calculated as the sum of van der Waals radii of 1.2 Å for aliphatic amide H and 1.5 Å for carbonyl 0. Since the entire electron shell of hydrogen is appreciably shifted onto the atom to which hydrogen is covalently bound, the shell repulsion between contact partners is small, and the attracting charges can approach each other more closely. Such short distance approach gives rise to a high attractive Coulomb energy as shown in Equation 4 and, therefore, a high dispersion energy. Thus, hydrogen bonds can be regarded as predominantly electrostatic interactions.[5] The resulting interaction energy (3 kcal/mol) is intermediate between the energies of van der Waals contacts (0.03 kcal/mol), and covalent bonds (7 kcal/mol).

III. HYDROPHOBIC INTERACTION

There are two schools of thought which exist for interpreting hydrophobic phenomena of proteins. The one is to explain hydrophobic interaction as entropic forces,[6] while the other is to regard it as a special case of van der Waals attraction when the solute molecules are placed in solvents.[7] In the following discussion each theory is explained separately.

A. Entropic Forces

The hydrophobic effect arises from the unfavorable interactions between water molecules and the nonpolar residues of a protein. When a hydrocarbon molecule is introduced into water, it induces changes in water structure, frequently decreasing its entropy. To minimize this unfavorable entropy change, the nonpolar molecules are forced to coalesce together into droplets or globules, reducing their surface of contact with water. As a result, the protein chain is forced to fold into a micellar structure with the hydrocarbon moiety on the inside of the globule and the polar groups on the outside. The removal of the nonpolar residues from contact with water makes a major contribution to the free energy of conformational stabilization. Thus, in this theory, the hydrophobic effect is purely the result of the phobia of water from contact with hydrocarbons. It is not due to the mutual affinity between the nonpolar chains through van der Waals forces, as is sometimes implied.[6] This is a complete contradiction to the other theory of the hydrophobic effect, as described in the following section, B. "Solvent Effects".

The early views of solutions, postulating increased water structure around the monomeric solute, were based on the interpretation of the unusual thermodynamic parameters for dissolving nonpolar solutes in water, namely a large negative entropy change and a small enthalpy change.

It is worth noting that the enthalpy and free energy of transfer of phenol into water from octane and from toluene are considerably different. It has been argued that changes of water structure do not play a major role in the solution process of hydrocarbons because the major component of the free energy of transfer stems from van der Waals interactions between the nonpolar groups in the nonaqueous solution. Van der Waals interactions, therefore, may be important in both solutions. On the other hand, these criticisms disregard the unusually large changes in entropy and heat capacity that are observed to accompany the transfer of nonpolar solutes to water. Van der Waals interactions alone cannot explain these changes, instead, structural changes in water must be invoked.[6]

According to Ross and Subramanian,[8] often the thermodynamic parameters characterizing self-association and ligand binding of proteins at 25°C, i.e., $\Delta G°$, $\Delta H°$, $\Delta S°$, and $\Delta C_p°$, are all of negative sign. The polarizability of a ligand in binding to a protein contributes to large negative values for the thermodynamic parameters.

The following conclusions have been drawn from thermodynamic study: (1) the only contributions to positive entropy and enthalpy changes arise from ionic and hydrophobic interactions; and (2) the only sources of negative enthalpy and entropy changes are van der Waals interactions and hydrogen bond formation in low dielectric media and protonation accompanying association.

The majority of researchers first compute the loss in translational and rotational entropies upon forming an association complex. The result is an unfavorable contribution of between 20 and 30 kcal/mol to the free energy of association at 27°C. The result of the hydrophobic portion of their calculation is a negative (favorable) contribution to the association free energy that outweighs the positive (unfavorable) contribution arising from the translational rotational immobilization.[8]

The criteria of Kauzmann[9] for hydrophobic interactions ($\Delta G° < 0, \Delta H° \simeq 0$, $\Delta S° > 0$, and $\Delta C_p° < 0$) are generally satisfied for $\Delta G°$ and $\Delta C_p°$, but in most instances are not satisfied for the changes in enthalpy and entropy, $\Delta H°$ and $\Delta°$. This contradiction between negative values of $\Delta H°$ and $\Delta S°$ and the positive values expected for hydrophobic interactions has been noted previously for several protein association processes. It is not possible to account for the stability of association complexes of proteins on the basis of hydrophobic interactions alone.[8]

There is little net energy difference when an amino acid forms a hydrogen bond with another group in an aqueous environment. However, an important source of negative contribution to $\Delta H°$ and $\Delta S°$ will arise if a hydrogen bond is formed in an environment of low dielectric constant. Typically such processes result in considerable changes in enthalpy and entropy of $\Delta H° = -5$ kcal/mol and $\Delta S° = -10$ to -20 cal/°K.mol.

Thus formation of several hydrogen bonds in a low dielectric environment, such as the areas between proteins which are inaccessible to water or the ligand-binding sites in the interior of an enzyme, could collectively make substantial negative contributions to $\Delta H°$ and $\Delta S°$.

Interactions between ionic species in aqueous solution are characterized by extremely small enthalpy changes and positive entropy changes. The release of the Bohr protons are estimated at $\Delta H° = 11$ kcal/mol proton. It is not possible to make a quantitative estimate of the energetics of proton-transfer contributions in this example. While the energetics of proton release or uptake are incidental in protein association processes, in general, they may make only minor contributions to the overall energetics.

The enhancement of solvent structure by the dissolution of nonpolar groups in water is greatest at low temperatures, and the tendency for hydrophobic association is increased because of this. But at high temperatures, as the solvent structure is randomized, the positive contribution to $\Delta S°$ by hydrophobic association is diminished while the enthalpy term dominates the stability of the complex.

The contribution to $\Delta C_p°$ from the hydrophobic effect and changes in soft vibrational modes are both negative, producing the observed large negative values of $\Delta C_p°$. As for the entropy, the hydrophobic contribution is positive while the vibrational contribution is negative. With increasing temperatures, the vibrational contribution becomes more negative but the positive hydrophobic contribution is diminished, thus making the $\Delta S°$ more negative or less positive. Both approaches stress the importance of the diminishing solvent structure at high temperatures resulting in the elimination of positive $\Delta S°$ due to hydrophobic interaction.

The tendency for protein association reactions to become entropy dominated and/or entropy-enthalpy assisted at low temperatures and enthalpy dominated at high temperatures arises from the diminution of the hydrophobic effect with increasing temperature, which is a general property of the solvent, water.

B. Solvent Effects

The chemical potential μ_1 of a dilute solution of macromolecules is expressed as:

$$\mu_1 = \mu_1^0 + RT \ln(\chi_1 \cdot f_1) \tag{6}$$

where R is the universal gas constant, T is the absolute temperature, μ_1^0 is the standard chemical potential of the solute, χ_1 is the solvent mole fraction, and $.f_1$ is the solvent activity coefficient.

The following relationship exists between solute and solvent:

$$\ln x_1 = \ln(1 - \chi_2) \simeq -\chi_2 \simeq -c_2 V^\circ_{m,1}/M_2 \tag{7}$$

where χ_2, c_2, and M_2 are the mole fraction, concentration, and relative molecular mass of the solute and $V^\circ_{m,1}$ is the molecular volume of pure solvent.

Furthermore, the solvent activity coefficient f_1 is defined as:

$$\ln f_1 = \sum_{i=2} a_i c_2^i \tag{8}$$

where a_i ($i \geq 2$) are unknown constants.

In their classical treatment of a polymer solution based on the lattice model, Flory[10] and Huggins[11] showed, independently, that the solvent chemical potential could be written in the form:

$$\mu_1 - \mu_1^0 = RT[\ln(1 - \phi_2) + (1 - m^{-1}) \phi_2 + \chi\phi_2^2] \tag{9}$$

where ϕ_2 is the volume fraction of solute, m is the molecular weight of the solute, and χ is a molecular parameter (Flory-Huggins parameter) defined by:

$$\chi = (2\epsilon_{12} - \epsilon_{11} - \epsilon_{22}) \, z/2kT \tag{10}$$

where k is Boltzmann's constant, z is the lattice coordination number, and ϵ_{11}, ϵ_{22}, and ϵ_{12} are the attractive energies between two solvent molecules, two solute segments, and a solvent molecule and a solute segment, respectively.

For the case $\chi = 0$, temperature has no effect on structure and the solvent is "athermal". In most macromolecular solutions, however, the Flory-Huggins parameter χ is positive because the energies ϵ_{ij} are mainly determined by van der Waals interactions which are approximately proportional to the product of molecular polarizabilities α_i and α_j, i.e.,:

$$\epsilon_{ij} = -K\alpha_i\alpha_j (i,j = 1,2) \tag{11}$$

where K is a positive constant.

A mixture of colloidal particles, macromolecules and solvent may become thermodynamically unstable if the temperature is lowered or the solvent composition is changed. The Flory-Huggins theory tells us that there is a critical χ value (χ_c) at which a plot of μ_1 vs. ϕ_2 at constant temperature and pressure shows a point of inflection. For $\chi < \chi_c$, the macromolecular solution is thermodynamically stable, but for $\chi > \chi_c$, it separates spontaneously into fluid phases of different composition.

The critical Flory-Huggins parameter can be calculated as:

$$\chi_c = (1 + m^{1/2})^2/2m \tag{12}$$

$$\phi_2 = 1/(1 + m^{1/2}) \tag{13}$$

For a solute of high molecular weight, χ_c approaches $1/2$, hence the $\chi = 1/2$ status becomes important in many polymer systems.

For two identical spheres of radius a, whose centers are a distance r apart, the total energy of interaction is found to be:

$$U_A(r) = \int_{r-a}^{r+a} U\rho_i(\pi x/r) \{a^2 - (r - x)^2\} \, dx \tag{14}$$

$$= (A_H/6)[\{2a^2/(r^2 - 4a^2)\} + (2a^2/r^2)$$

$$+ \ln\{1 - 4a^2/r^2\}] \tag{15}$$

The quantity A_H is named after Hamaker,[12] who showed that a sum of dispersion energies within an interaction range for colloidal particles is of the order of their size. The Hamaker coefficient depends on the density ρ and polarizability Λ of the materials:

$$A_H = \pi^2 \, \rho_i\rho_j \, \Lambda_{ij} \tag{16}$$

It is convenient to express U in terms of the surface-to-surface separation d:

$$U_A(d) = -(A_H a/12d)[1 + \{3/8 + \ln(d/a)\} (2d/a)] \quad (d \ll a)$$

$$= -16A_H a^6/9d^6 \quad (d \ll a) \tag{17}$$

Optimally defined Hamaker coefficient is:

$$A_H = 3/4h\nu_o\pi^2 \, \rho^2\alpha_o \tag{18}$$

where h is Planck's constant, α_o is the static polarizability, and ν_o is the UV absorption frequency. For solute particles 1 immersed in a solvent 2, the effective Hamaker coefficient is given by:

$$A_H = A_{11} + A_{22} - 2A_{12} \tag{19}$$

where each coefficient A_{ij} is calculated from optical dispersion data in the dilute gas limit. A similarity is apparent between this equation and Equation 10.

If A_{12} is the geometric mean of A_{11} and A_{22}, then:

$$A_H = (A_{11}^{1/2} - A_{22}^{1/2})^2 \tag{20}$$

The net force due to dispersion forces is always attractive if the particles are made of the same material. The net forces between two particles made of different materials may be either positive or negative.

The potential of the van der Waals interaction between bodies of materials 1 and 2 across a medium 3, is given by the product of a geometric term that is influenced by the size and shape of each body, and a Hamaker coefficient A_{132}:

$$A_{132} = A_{12} + A_{33} - A_{13} - A_{23} \tag{21}$$

where A_{ij} is the Hamaker coefficient for the van der Waals attraction of materials i and j across a vacuum.

For the interaction of two particles of the same solute 1 across a solvent 3, the relation is:

$$A_{131} = (\sqrt{A_{11}} - \sqrt{A_{33}})^2 + 3/4 \; kT(\epsilon_1 - \epsilon_3/\epsilon_1 + \epsilon_3)^2 \tag{22}$$

The contribution from dispersion and from interactions other than dispersion to the surface tension of water are approximately 22 dyn/cm and 50 dyn/cm, respectively. Water can interact with nonpolar materials only by dispersion. Consequently, the contribution from orientation and induction to A_{13} and A_{23} would be negligible. Dispersion will make a significant contribution to A_{131} or A_{132} for the interaction of nonpolar materials in water, as can be checked by setting $A_{11} > A_{33}$ in Equation 22.

In a medium less polar than water, orientation and induction occur to a lesser extent than in water. Hence, the contribution to A_{132} or A_{131} in a nonpolar medium would be small as $\epsilon_1 \simeq \epsilon_3$ in Equation 22. Besides, the difference between A_{11} or A_{12} and A_{33} will be less for the interaction of nonpolar materials when the medium is nonpolar than when the medium is water. Again, from Equation 22 it can be seen that the dispersion contribution to A_{131} (or A_{132}) will be smaller in a nonpolar medium, compared with an aqueous one.

In total, two nonpolar materials will undergo a larger van der Waals attraction with each other when water is the intervening medium than when the medium is nonpolar.

The Hamaker coefficient for the interaction of dissimilar materials can become negative when the strength of the van der Waals interactions among the molecules of the medium lies between those of interactions among the molecules of each material.

The van der Waals interaction between the two different solutes 1 and 2, that are immersed in solvent 3, will be repulsive when the Hamaker coefficient A_{132} is negative in the expression for the free energy ΔF of the interaction:

$$\Delta F = -A_{132}/12\pi d^2 \tag{23}$$

where d is the distance between two particles and/or molecules, taken to be semi-infinite homogeneous slabs. It is easily shown that this is always the case when the individual Hamaker coefficients of the two solutes, A_{11} and A_{22}, have values that straddle that of the solvent, A_{33}:

$$A_{11} > A_{33} > A_{22} \quad \text{or} \quad A_{11} < A_{33} < A_{22} \tag{24}$$

Thus, in solvent 3, solutes 1 and 2 will undergo a van der Waals repulsion when the interfacial free energy of solute 1 is smaller and that of solute 2 is larger than that of the solvent, or vice versa.

The ''hydrophobic effect'' is an attraction in aqueous medium between macromolecules, by which some of the more hydrophobic sites, upon approaching one another, preferentially interact with each other, frequently by van der Waals interactions, which tend to be attractive between low surface energy materials immersed in high energy solvents such as water:

$$A_{33} > A_{11} \quad \text{and} \quad A_{33} > A_{22} \tag{25}$$

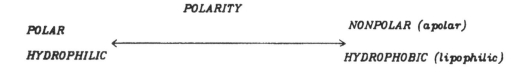

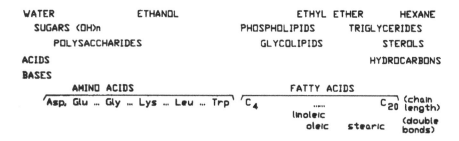

FIGURE 1.   Relative polarity of important food components.

The traditional need to invoke "anomalous" forces to explain special colloidal effects, e.g., Stern layer, structural water, and hydrophobic interaction, etc., is unnecessary if statistical factors are properly induced into van der Waals forces.

C. Discussion

Whether "hydrophobic effect" can be considered as one of the independent noncovalent forces is uncertain. Van der Waals forces include a variety of component forces, i.e., electron shell repulsion, dispersion forces, and electrostatic interactions, and the dispersion forces contain long-range force, orientation force and induction force. The extents of involvement of these component forces are different depending on inter- and intra-molecular or interatomic orientation. As discussed before, if orientation and induction forces are the major components of the hydrogen bond, then van der Waals potentials can be regarded as the principal forces, and other noncovalent forces can be interpreted as alterations of the van der Waals forces. It is then acceptable that the hydrophobic effect is also a special expression of the van der Waals potentials. It may be feasible, therefore, that a great many phenomena, e.g., the behavior, including chromatographic separation, and even the function of proteins can be explained by this unified force. However, it is premature to prove that this is feasible. Explanation of thermodynamic behavior of proteins, especially by using the solvent effects of van der Waals forces is imminent.

IV. HYDROPHOBICITY, POLARITY, AND SOLUBILITY

There is a well-known general rule with regard to solubility, i.e., "like dissolves like". This rule means that a compound (solute) which is similar in structure and chemical/physical properties to a solvent, dissolves well in that solvent. If the compound is a solvent, these two solvents are miscible in one another. The above rule can be re-worded as, "If the polarities match, the solute easily dissolves in the solvent." "Polarity" is, therefore, a scale generated in an effort to classify compounds based on the "mutual miscibility". The adjectives "hydrophobic" (or "lipophilic") and "hydrophilic", have been used as synonyms of "nonpolar" (or "apolar") and "polar", respectively (Figure 1).

HLB (hydrophile lipophile balance) scale for the emulsification of lipids with emulsifiers is based on a similar concept to that of solvation of solutes in solvents. In the case of emulsification, the HLB value of a surfactant should match the HLB value of the lipid (required HLB) for forming the most stable emulsion. Separation of compounds by a variety of chromatographic techniques can also be explained by differences in polarity. Thus, the phenomena of protein hydrophobicity, lipid emulsification, and chromatography are all based on a similar theoretical background. Therefore, the use of hydrophobicity relationships is not restricted only to proteins and their relating compounds, rather it can be expanded broadly to other food components.

A. Polarity and its Solvent Effects

The rule "like dissolves like" should not be narrowly interpreted, especially in terms of chemical structure of compounds. There are many examples of solutions of chemically dissimilar compounds. For example, the solvent pairs of methanol and benzene, water and N,N-dimethylformamide, aniline and diethylether, or polystyrene and chloroform, are completely miscible at room temperature. Conversely, insolubility can occur in spite of similarity of the two partners. Thus, polyvinylalcohol does not dissolve in ethanol, acetyl cellulose is insoluble in ethyl acetate, and polyacrylonitrile is insoluble in acrylonitrile. Between these two extremes there is a whole range of possibilities where the two materials dissolve each other to a limited extent.[13]

Solvation effects depend on the intermolecular interaction between solvent and solute molecules, which determines the mutual solubility. A compound A dissolves in a solvent B only when the homo-intermolecular forces of attraction K_{AA} and K_{BB} for the pure compounds can be overcome by the hetero-forces K_{AB} in solution. The sum of the interaction forces between the molecules of solvent and solute can be related to the polarity of A and B.

Since the dielectric constant ϵ and dipole moment μ are both important complementary solvent properties, it has been recommended that solvents are classified according to their electrostatic factor EF (defined as the product of ϵ and μ), which takes into account the influence of both properties.

In the literature, dielectric constants as well as dipole moments are often used in the quantitative characterization of solvent polarity. However, the characterization of a solvent by its "polarity" is an unsolved problem since the term "polarity" itself has never been precisely defined. Usage of the term "polarity" can imply: (1) the permanent dipole moment of a compound (2) its dielectric constant or (3) the sum of all those molecular properties responsible for all of the interaction forces between solvent and solute molecules. As far as solvents are concerned, the polarity is the "overall solvation ability".

B. Empirical Parameters of Solvent Polarity

Because of the complex nature of solvation phenomena, no single macroscopic physical parameter such as dielectric constant could possibly account for the multitude of solute-solvent interactions on the molecular-microscopic level. Until now the complexity of solute-solvent interactions has also prevented the derivation of generally applicable mathematical expressions, which would allow the calculation of reaction rates or equilibrium constants of reactions carried out in solvents of different polarity.

In such a situation, other indices of solvent polarity have been sought. The lack of reliable theoretical expressions for calculating solvent effects and the inadequacy of defining "solvent polarity" in terms of simple physical constants have stimulated attempts to introduce empirical scales of solvent polarity, based on convenient, well-known, solvent-sensitive reference processes. A common approach is to assume that some particular reaction rate, equilibrium, or spectral absorption is a suitable model for a large class of solvent-dependent processes.

1. Empirical Parameters Based on Equilibria

The first attempt was made by Meyer[14] using keto-enol tautomerism of β-dicarbonyl compounds. The tautomeric equilibrium constant K_T is defined as:

$$K_T = [enol]/[diketo] = L \cdot E \tag{26}$$

where E is the enol constant and measures the enolization capacity of the keto form and L is the demotropic constant which is a measure of the enolization power of the solvent. Since E = 1 for ethyl acetoacetate by definition, the values of L are equal to the equilibrium constants of ethyl acetoacetate being used as empirical parameter of solvent polarity.

Eliel and Hofer[15] used 2-isopropyl-5-methoxy-1,3-dioxan which changes to the more dipolar axial *cis* isomer in polar solvents. The authors proposed to use "D_1-scale" which was the standard free energy changes $\Delta G°_{OCH3}$

Gutmann[16] proposed the "Donor Number" DN_{SbCl_5} which was calorimetrically measured as $-\Delta H_{D \cdot SbCl_5}$ (kcal/mol), where $D - SbCl_5$ is the adduct formed between antimony pentachloride and electron-pair donor solvent D. However, since the electron-pair donor (EPD)-acceptor (EPA) interactions are only one of many solute-solvent interactions, solvent donicity based on the Lewis basicity of solvents may lack versatility for use as a polarity scale.

The solvent-dependent tautomerism of a pyridoxal 5'-phosphate Schiff's base has been shown to be another appropriate model process for the measurement of solvent polarities.[17] This model process seems to be particularly useful for the determination of the polarities of sites of proteins at which pyridoxal 5'-phosphate is bound.

2. Empirical Parameters Based on Kinetics

Since reaction rates can be strongly affected by solvent polarity, the reaction rates of selected chemical reactions have been used as an index of solvent polarity. However, the major drawback of this approach is in the restrictive solubilization in the broad range of polarity difference for different solvents.

Winstein and Fainberg et al.[18] found that the S_N1 solvolysis of tert-butylchloride was strongly accelerated by polar, especially protic, solvents. Grunwald and Winstein[19] defined a solvent "ionizing power" parameter Y as:

$$Y = \log k^t_{A - BuCl} - \log k^t_{0 - BuCl} \tag{27}$$

where $k^t_{0 - BuCl}$ is the first-order rate constant for the solvolysis of *tert*-butyl chloride at 25°C in aqueous ethanol as reference (80% ethanol; Y = 0), and $k^t_{A - BuCl}$ is the corresponding rate constant in another solvent. Ionization of the $C - Cl$ bond in $t - BuCl$ is the rate determining step. Due to the fact that the reaction rate depends not only on the ionizing power of the solvent but also on the nucleophilicity of the solvent, Winstein et al.[20] later provided a modified equation:

$$\log k_A - \log k_0 = m \cdot Y + 1 \cdot N \tag{28}$$

where m and l are substrate parameters, Y is the solvent ionizing power, and N measures solvent nucleophilicity.

Drougard and Decroocq[21] suggested that the value of log k for the S_N2 Menschutkin reaction of tri-*n*-propylamine and methyl iodide at 20°C be termed "φ" as follows:

$$\varphi = \log k_2[(CH_3CH_2CH_2)_3 N + CH_3I] \tag{29}$$

Gielen and Nasielski[22] suggested that a solvent polarity scale could be based on an electrophilic aliphatic substitution reaction such as the reaction of bromine with tetramethyltin. X-values are defined as:

$$\log k_A - \log k_0 = p \cdot X \tag{30}$$

where k_A and k_O are the rate constants of the reaction in a given solvent A and in glacial acetic acid, respectively, and p is a parameter characteristic of the given reaction. The two solvent scales, X and Y, do not quite agree, since the reaction rate is further influenced by the electrophilic and nucleophilic characters of the solvent.

Berson et al.[23] used the fact that the rate of some Diels-Alder cycloaddition reactions is affected by the solvent. In the Diels-Alder addition of cyclopentadiene to methyl acrylate, the ratio of the endo-product to the exo-product depends on the reaction solvent. The endo-addition is favored with increasing solvent polarity. Berson et al.[23] defined:

$$\Omega = \log[endo]/[exo] \tag{31}$$

Owing to the low solubility of the reactants in polar media, an extension of this scale is limited.

3. Empirical Parameters Based on Spectroscopic Measurements

Kosower[24] used the longest wavelength intermolecular charge-transfer transition of 1-ethyl-4-methoxycarbonylpyridinium iodide as a model process. A solvent change from pyridine to methanol causes a hypsochromic shift of the charge-transfer band of 105 nm. This is due to stabilization of the electronic ground state (which is an ion pair) relative to the first excited state (which is a radical pair) with increasing solvent polarity. Mohammed and Kosower[25] defined their polarity parameter, Z, as the molecular transition energy, E_T (kcal/mol) according to :

$$E_T = h \cdot c \cdot \bar{v} \cdot N$$

$$= 2.859 \cdot 10^{-3} \cdot \bar{v} \tag{32}$$

where h is Planck's constant, c is the velocity of light, $\bar{v}$ is the wavenumber of the photon which produces the electronic excitation, and N is Avogadro's number.

Dimroth et al.[26] have proposed a solvent polarity parameter, $E_T(30)$, based on the transition energy for the longest-wavelength solvatochromic absorption band of the pyridinium-N-phenoxide betaine dye (dye no. 30). According to Equation 32, the $E_T(30)$ value for a solvent is the transition energy of the dissolved betaine dye no. 30. The major advantage of this method is that the solvatochromic absorption band is at longer wavelengths than for Kosower's dye, generating an extraordinarily large range for the solvatochromic behavior, for example, $\lambda = 810$ nm, $E_T(30) = 35.5$ for diethylether and $\lambda = 453$ nm, $E_T(30) = 63.1$ for water. The major limitation of the $E_T(30)$ values is the fact that no $E_T(30)$ values can be measured for acidic solvents such as carboxylic acids. Addition of traces of an acid to solutions of dye immediately changes the color to pale yellow due to protonation at the phenolic oxygen of the dye.

As shown in Table 1, solvents such as formamide ($\epsilon = 111.0$) and N-methylformamide ($\epsilon = 182.4$) which are described as highly polar on account of their dielectric constants (ϵ), are by no means as polar as believed in their behavior towards the betaine dye ($E_T(30) = 56.6$ and 54.1, respectively). Similarly, N-methylacetamide, the solvent with the highest dielectric constant ($\epsilon = 191.3$), is in fact no more polar than ethanol according to their $E_T(30)$ values of 52.0 and 51.9, respectively.

Table 1
EMPIRICAL PARAMETER OF SOLVENT POLARITY, $E_T(30)$
kcal/mol[a]

Solvent	$E_T(30)$	Solvent	$E_T(30)$
Water	63.1	Pentanone	41.1
Tetrafluoropropanol	59.4	Dichloromethane	41.1
Glycerol	57.0	Tetramethylurea	41.0
Formamide	56.6	Hexamethylphosphoric triamide	40.9
Ethanediol	56.3	Cyclohexanone	40.8
Methanol	55.5	Pyridine	40.2
Propanediol	54.9	Hexanone	40.1
Methylformamide	54.1	Methyl acetate	40.0
Diethylene glycol	53.8	Dichloroethane	39.4
Ethanol/Water (80:20)	53.7	Quinoline	39.4
Triethylene glycol	53.5	Pentanone	39.3
Methoxyethanol	52.3	Tetramethylguanidine	39.3
Methylacetamide	52.0	Chloroform	39.1
Ethanol	51.9	Heptanone	38.9
Aminoethanol	51.8	Triethyleneglycol dimethylether	38.9
Acetic acid	51.2	Methylpyridine	38.3
Benzyl alcohol	50.8	Fluorobenzene	38.1
1-Propanol	50.7	Ethyl acetate	38.1
1-Butanol	50.2	Iodobenzene	37.9
Isobutanol	49.0	Bromoethane	37.6
2-Propanol	48.6	Bromobenzene	37.5
Cyclopentanol	47.7	Chlorobenzene	37.5
Dimethylphenol	47.6	Tetrahydrofuran	37.4
2-Butanol	47.1	Chloropropane	37.4
Isoamyl alcohol	47.0	Anisole	37.2
Cyclohexanol	46.9	Dichlorobenzene	37.0
Propylene carbonate	46.6	Phenetole	36.4
2-Pentanol	46.5	Diethyl carbonate	36.2
Nitromethane	46.3	Trichloroethane	36.2
Acetonitrile	46.0	Dioxane	36.0
3-Pentanol	45.7	Trichloroethylene	35.9
Dimethylsulfoxide	45.0	Piperidine	35.5
Aniline	44.3	Diethylamine	35.4
Tetrahydrothiophenedioxide	44.0	Diphenyl ether	35.3
tert-Butanol	43.9	Diethyl ether	34.6
Acetic anhydride	43.9	Benzene	34.5
Dimethylformamide	43.8	Diisopropyl ether	34.0
Dimethylacetamide	43.7	Toluene	33.9
Propionitrile	43.7	*tert*-Butylbenzene	33.7
Nitroethane	43.6	Dibutyl ether	33.4
Trimethyl phosphate	43.6	Diisopropylamine	33.3
Methylpyrrolidinone	42.2	Triethylamine	33.3
Acetone	42.0	Xylene	33.2
Benzonitrile	42.0	Mesitylene	33.1
Nitrobenzene	42.0	Carbon disulfide	32.6
Diaminoethane	42.0	Carbon tetrachloride	32.5
Methyl butanol	41.9	Tetrachloroethylene	31.9
Butanone	41.3	Cyclohexane	31.2
Acetophenone	41.3	Hexane	30.9

[a] Adapted from Reichardt[13]

The $E_T(30)$ values of binary solvent mixtures are not linearly related to their composition. Addition of a small amount of a polar solvent to a solution of the dye in nonpolar solvents causes a nonlinear hypochromic shift, indicating solvation of the dipolar betaine molecule by the more polar solvent.

The longest-wavelength solvatochromic absorption is affected by changes in temperature and pressure. The lower the temperature and the higher the pressure, the higher the $E_T(30)$ value. However, this conclusion must be carefully interpreted, as other factors such as conformational changes may also contribute to the observed spectral changes.

Kamlet et al.[27] introduced a π^* scale of solvent polarities. This scale is so named because, it is derived from solvent effects on the $p{\to}\pi^*$ and $\pi \to \pi^*$ electronic spectral transitions of seven nitroaromatics compounds. Solvent effects on the $\bar{v}_{max}$ values (maximum wave number) of these seven indicators were employed in the initial construction of the π^* scale, which was then expanded and refined by multiple least-squares correlations with 40 additional solvatochromic indicators. In this way, an averaged scale of solvent polarities, comprising seventy solvents, has been established.

4. Empirical Parameters Based on Chromatographic Behavior

In gas-liquid chromatography it is possible to determine the solvent power of the stationary liquid phase very accurately for a large number of substances. The retention index (I) was defined by Kovats[28] as:

$$I = 100 \cdot (\log V_x - \log V_n)/(\log V_{n+1} - \log V_n) + 100 \cdot n \qquad (33)$$

where V_x, V_n, and V_{n+1} are the corrected retention volumes of the test substance and of the standard n-alkanes containing n and $n + 1$ carbon atoms, respectively ($V_n \leqslant V_x \leqslant V_{n+1}$). The retention index is independent of the gas chromatographic equipment used, and depends only on the test substance, temperature, and stationary phase.

Kovats and Weisz[29] used these retention indices to examine the polarity of stationary liquid phases. They calculated the difference between the retention indices of the test substance on the column with a given phase x and with a standard nonpolar stationary phase S at temperature T. This difference, $\Delta I_T^x = I_T^x - I_T^s$, was then proposed as the solvent polarity parameter. Using 1-chloro- and 1-bromo-n-hexadecane as the standard dipolar stationary phase, and n-hexadecane as the standard nonpolar stationary phase, $\Delta I_{50}^{Cl/Br}$ was computed as a new solvent polarity parameter. This value showed correlations with other empirical polarity parameters such as Z, $E_T(30)$, and log k_1 of p-methoxyneophyl tosyl solvolysis.

Another approach was proposed by Rohrschneider[30] which depended not only on the nature of the stationary phase but also on the type of substance analysed:

$$\Delta I = a \cdot x + b \cdot y + c \cdot z + d \cdot u + e \cdot s \qquad (34)$$

where a, b, c, d, and e were polarity factors of the stationary liquid phase, and x, y, z, u, and s were polarity factors of the substances undergoing analysis. Having determined the polarity factors of 22 different stationary phases with respect to chosen standards (benzene, ethanol, 2-butanone, nitromethane, and pyridine), and those of 30 substances, Rohrschneider calculated 660 retention indices with only small error. These tabulated retention data allow the calculation of retention indices of many other substances.

Similar approaches were employed by Snyder[31] to the measurement of solvent polarity for liquid chromatography. To compute the polarity, the experimental solute distribution coefficients K_g were corrected for solvent molar volume V_s as: $K_g' = K_g \cdot V_s$. n-Octane (o), ethanol (e), dioxane (d), and nitromethane (n) were used as the test solutes. These adjusted distribution coefficients K_g' were then used to calculate polar distribution coefficients K_g'':

Table 2
EMPIRICAL PARAMETER OF SOLVENT POLARITY, P'[a]

Solvent	P'	Solvent	P'
Hexane	0.1	Cyclohexanone	4.7
Iso-octane	0.1	Dioxane	4.8
Cyclohexane	0.2	Benzonitrile	4.8
Carbon disulfide	0.3	Acetophenone	4.8
Decane	0.4	Picoline	4.9
Squalane	1.2	Quinoline	5.0
Carbon tetrachloride	1.6	Acetone	5.1
Triethyl amine	1.9	Methanol	5.1
Butyl ether	2.1	Nitroethane	5.2
Isopropyl ether	2.4	Diethylene glycol	5.2
Toluene	2.4	Pyridine	5.3
para-Xylene	2.5	Methoxy ethanol	5.5
Benzene	2.7	Cyano morpholine	5.5
Chlorobenzene	2.7	Triethylene glycol	5.6
Bromobenzene	2.7	Benzyl alcohol	5.7
Iodobenzene	2.8	Acetonitrile	5.8
Ethyl ether	2.8	Tetramethyl urea	6.0
Methylene chloride	3.1	Acetic acid	6.0
Fluorobenzene	3.2	Methyl formamide	6.0
Ethoxybenzene	3.3	Nitromethane	6.0
Octanol	3.4	Tetramethyl guanidine	6.1
Phenyl ether	3.4	Propyl carbonate	6.1
Ethylene chloride	3.5	Aniline	6.3
Isopentanol	3.7	Dimethyl formamide	6.4
Anisole	3.8	Formyl morpholine	6.4
Butanol	3.9	Dimethyl acetamide	6.5
Isopropanol	3.9	Butyrolactone	6.5
Tetrahydrofuran	4.0	tris-Cyanoethoxypropane	6.6
Propanol	4.0	Methyl-pyrrolidone	6.7
Chloroform	4.1	Oxydipropionitrile	6.8
tert-Butanol	4.1	Tetrahydrothiophene-dioxide	6.9
Benzyl ether	4.1	Ethylene glycol	6.9
Ethanol	4.3	Dimethyl sulfoxide	7.2
Ethyl acetate	4.4	Hexamethyl phosphoric triamide	7.4
Nitrobenzene	4.4	meta-Cresol	7.4
Lutidine	4.5	Tetrafluoropropanol	8.6
bis-Ethoxy ethyl ether	4.6	Dodecafluoroheptanol	8.8
Tricresyl phosphate	4.6	Formamide	9.6
Methylethyl ketone	4.7	Water	10.2

[a] Adapted from Snyder.[31]

$$\log K_g^{''} = \log K_g^{'} - \log K_v \tag{35}$$

where K_v is the $K_g{}'$ value of the hypothetical n-alkane with the same molar volume V_x as the solute X. K_v is in turn calculated from the $K_g{}'$ value of n-octane (K_o):

$$\log K_v = (V_x/103)(\log K_o) \tag{36}$$

Finally, P' is calculated as the sum of $\log K_g^{''}$ values for the three solutes ethanol, dioxane, and nitromethane (Table 2). As there is a relationship: $\log K_g^{''} = (b/100)\Delta I$, where b is the logarithm of the retention-time increment due to addition of a methylene group to the solute molecule, Snyder's P' value is closely related to Rohrschneider's I.

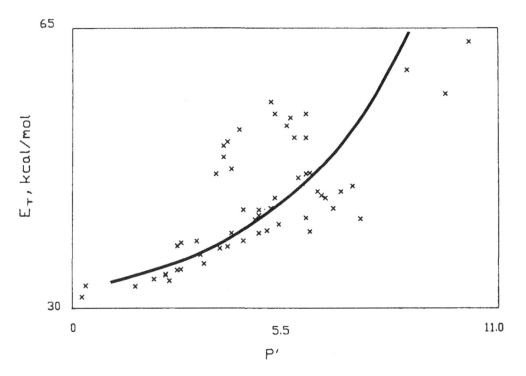

FIGURE 2.   Relationship between the two most popular polarity scales, E_T (30) and P'.

According to Abboud et al.,[32] among the eight most widely used polarity scales, there exist good correlations with r = 0.97 to 0.99 (n = 28, p < 0.01); e.g., between π^* and E_T, r = 0.987. However, with 57 solvents, r = 0.74 was obtained between E_T(30) and P' (Figure 2), which was improved to r = 0.80 when analyzed for log E_T(30) vs. P'.

V. EFFECTS OF SALTS AND UREA ON HYDROPHOBICITY

A. Effects of Salts on Protein Hydrophobicity

According to Melander and Horvath[33] the solubility S of proteins in salt solutions of molality m can be expressed relative to the solubility S_0 in pure water as:

$$\ln(S/S_0) = \Delta G_{e.s.}^{net} (m) - N\phi/RT\sigma m$$

where $\Delta G_{e.s.}^{net}$ is the net energy change due to electrostatic interactions, ϕ is the nonpolar surface area per protein molecule which is also the reduction in the molecular surface area of the protein upon aggregation, and σ is the molal surface tension increment of the salt. The first term is the electrostatic interaction term, which is a function of salt concentration. This can be expressed as $\beta + \Lambda m$ where $\Lambda = D\mu/RT$, in which μ is the dipole moment of the protein and D is a constant. Figure 3 schematically illustrates the normalized solubility of proteins $\ln(S/S_0)$ as a function of the salt concentration [m], which is the net result of two antagonistic effects. This relationship is simply included in $K_s = \Omega\sigma - \Lambda$ when $K_s = -d\ln S/dm$.

Salts have been traditionally arranged in the Hofmeister (lyotropic) series according to their ability to precipitate proteins. The effect of salts on the polymerization of protein aggregates can be quantitatively described in terms of the molal surface tension increments (Table 3). Melander and Horvath[33] proposed that the molal surface tension of salts quan-

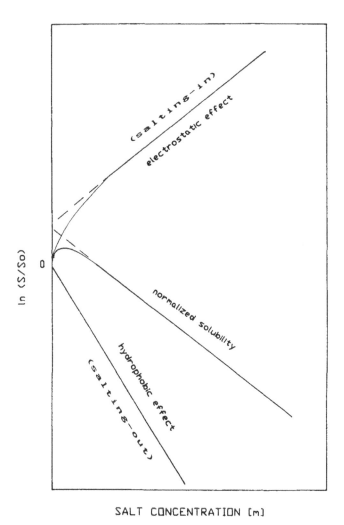

FIGURE 3. Scheme of the normalized solubility of proteins as a function
of the salt concentration. Adapted from Melander and Horvath.[33]

titatively represented the originally empirical lyotropic series which regulated association
and dissociation of protein molecules.

In the relationship $K_s = \Omega\sigma - \Lambda$, Ω relates to protein hydrophobicity through ϕ and Λ,
which is the effect of salts on the electrostatic free energy of the proteins, and is relatively
independent of the salt type or the pH. Therefore, change in K_s due to the change in salt
type is strongly affected by the size of σ value. Since the surface tension of water is mainly
caused by hydrogen bonding between water molecules at air-water interface (surface), positive σ entails the extent of intensifying the water-water hydrogen bonding at the surface.

The strong electric field of cations, along with their tendency to associate with the oxygen
atoms of water molecules, results in the normal structure of water being broken in the
presence of cations. The three-dimensional structures that result from the strong orientation
of water molecules on the cations, however, form a structure that is more organized than
the normal water structure. Anions readily associate with the available hydrogen atoms of
normal water assemblies. The anion-associated structures have their oxygen atoms outward.
Cations and anions together form multi-ionic assemblies in water which are more hydrogen
bonded than the structures in pure water. Therefore, the solution is held together by stronger

Table 3
MOLAL SURFACE TENSION
INCREMENTS OF SALTS,
δ (x 10 dyn g/cm mol)[a]

Salt	δ	Salt	δ
KSCN	0.45	K_2-tartrate	1.96
$NaClO_3$	0.55	$Ba(NO_3)_2$	2.0
NH_4I	0.74	LiF	2.0
LiI	0.79	Na_2HPO_4	2.02
KI	0.84	$NiSO_4$	2.10
NH_4NO_3	0.85	$MgSO_4$	2.10
KClO	0.86	$MnSO_4$	2.10
NaI	1.02	$CuSO_4$	2.15
$NaNO_3$	1.06	$(NH_4)_2SO_4$	2.16
NH_4Br	1.14	$ZnSO_4$	2.27
$LiNO_3$	1.16	Na_2-tartrate	2.35
LiBr	1.26	K_2SO_4	2.58
KBr	1.31	Na_3PO_4	2.66
NaBr	1.32	Na_2SO_4	2.73
CsI	1.39	Li_2SO_4	2.78
NH_4Cl	1.39	$FeCl_3$	2.78
$KClO_4$	1.4	$BaCl_2$	2.93
$FeSO_4$	1.55	K_3-citrate	3.12
LiCl	1.63	$MgCl_2$	3.16
NaCl	1.64	$CaCl_2$	3.66
$CsNO_3$	1.67	$K_4Fe(CN)_6$	3.9
$CuSO_4$	1.82	$K_3Fe(CN)_6$	4.34

[a] Adapted from Melander and Horvath.[33]

orientation, induction, and dispersion forces than water, and hence, is characterized by higher surface tension and larger A_{33} value.

On the contrary, large anions, due to their size, break up the normal structure of water but are unable to form their own assemblies since their electric fields are weak. Structure breaking also occurs when the ions, e.g., guanidinium or thiocyanate, consistently promote two-dimensional assemblies of water molecules as dictated by resonance structures of the ions rather than three-dimensional assemblies.[34] Thus, the solutions of structure-breaking ions have a lower proportion of three-dimensional assemblies than pure water, and would be characterized by a smaller A_{33} and surface tension than water. The surface tension values of $3M$ $(NH_4)_2SO_4$, $3M$ NaCl, $1M$ tetraethylammonium chloride, and $4M$ guanidinium chloride are 78.4, 76.9, 57.0, and 51.0 dyn/cm, respectively, compared to 72.0 dyn/cm for water.

A_{13} or A_{23} also may be increased by dispersion interactions between the ions and the protein, or the ions and the adsorbent, if the ion has a higher dispersion coefficient (a quantity proportional to the product of the polarizability of the molecules and the number of molecules per unit volume) than water. It can be inferred that this effect may be important for structure-breaking ions, since in $A_{132} = A_{12} + A_{33} - A_{13} - A_{23}$, the increase in A_{13} or A_{23} results in a decrease in A_{132}. In the case of structure-making ions, the dispersion effects, which increase A_{13} or A_{23}, are negligible, but enhanced A_{33} causes an increase in A_{132} or A_{232}.

The increase in A_{132} entails an increase in the attractive force between protein and adsorbent across the medium in the case of hydrophobic interaction chromatography. In the case of a protein solution, an increase in A_{232} renders the protein less soluble in water due to association as a result of addition of structure-forming salts.[35]

A "salting-out" salt at high ionic strength will increase the van der Waals attraction between protein and adsorbent across the medium, or between protein molecules leading to association, while a "salting-in" salt will reduce it.

In Table 3, δ value of about 1.5 marks the dividing line that is in balance with the Debye term, with K_s being 0, although this also depends on the size of Ω. Below this critical value of δ, denoted δ_c, salts become the salting-in type with negative K_s values, whereas above δ_c, salts are the salting-out type with positive K_s values.

Remarkably, the lyotropic order of anions parallels a variety of chemical reactions, not only for the dissociation and solubilization of antigen-antibody complexes, denaturation of proteins, depolymerization of protein polymers, and thermal stability of DNA, but also for the solubility of riboflavin, adenine, 2-methylnaphthoquinone, benzene, oxygen, and hydrogen.[36] The term "chaotropic" is used for those ions which favor the transfer of apolar groups to water.

Damodaran and Kinsella[37] discussed the changes in electrostatic and hydrophobic forces in proteins caused by ions inducing conformational changes in proteins, thus affecting their functional behavior. Some examples of the validity of the lyotropic rule are solubility of heated soy protein isolate,[38] gluten solubility and consistency of flour dough,[39] and thermal stability of storage proteins from fababean measured by differential scanning calorimetry.[40]

B. Effects of Urea on Hydrophobic Interaction

Wetlaufer et al.[41] found that hydrocarbons of sufficient size, larger than ethane, were more soluble in aqueous urea than water. They proposed two mechanisms for this phenomenon: (1) an indirect mechanism, whereby urea alters the "structure" of water in a way that facilitates the solvation of a hydrocarbon with water and (2) a direct mechanism, whereby the hydrocarbons are solvated by both urea and water molecules.

The indirect mechanism has received considerable attention in the literature. Frank and Franks[42] proposed that the water around urea was less hydrogen bonded than bulk water, and many experimental results have been cited as showing that urea acts as a water "structure breaker". However, there have also been objections raised to the indirect mechanism, stemming from other experiments which indicate no significant effect of urea on water structure and, perhaps of more importance, a lack of correlation between the effect of a given solute on water structure and its ability to solubilize hydrocarbons.

Robinson and Jencks[43] and Roseman and Jencks[44] have proposed that the increased solubility of hydrocarbons in aqueous urea results primarily from a smaller free energy of cavity formation in the mixed solvent, resulting from the replacement of water by the larger urea molecule in the solvation region. They note that the water-hydrocarbon interfacial tension is reduced by addition of urea, although the surface tension of aqueous urea is slightly greater than that of pure water. This fact entitles urea to be a salting-in type compound based on the theory of Melander and Horvath.[33]

According to Kuharski and Rossky,[45] in agreement with Roseman and Jencks,[44] the presence of urea in the solvation region of the apolar sphere is found to weaken the water-water interactions in this region. Urea-water interactions are enhanced in the solvation region as compared to aqueous urea. Each urea molecule displaces several water molecules from the apolar solvation shell.

REFERENCES

1. **London, F.,** Uber einige Eigenshaften und Anwendungen der Molekularkrafte, *Z. Phys. Chem.,* B11, 222, 1930.
2. **Van der Waals, J.,** Contribution to the theory of binary mixtures, *Proc. Acad. Amsterdam,* 10, 123, 1909.

3. **Jones, J. E.,** On the determination of molecular fields, *Proc. R. Soc. London, Ser. A,* 106, 441, 1924.
4. **Lifson, S.,** Molecular forces, in *Protein-Protein Interactions,* Jaenike, R. and Helmreich, E., Eds., Springer-Verlag, New York, 1972.
5. **Schulz, G. E. and Schirmer, R. H.,** in *Principles of Protein Structure,* Schulz, G. E. and Schirmer, R. H., Eds., Springer-Verlag, New York, 1979.
6. **Tanford, C.,** *The Hydrophobic Effect: Formation of Micells and Biological Membranes,* 2nd ed., Tanford, C., Ed., John Wiley & Sons, New York, 1980.
7. **Van Oss, C. J., Good, R. J., and Chaudhury, M. K.,** The role of van der Waal forces and hydrogen bonds in "hydrophobic interactions" between biopolymers and low energy surfaces, *J. Colloid Interf. Sci.,* 111, 378, 1986.
8. **Ross, P. D. and Subramanian, S.,** Thermodynamics of protein association reactions: forces contributing to stability, *Biochemistry,* 20, 3096, 1981.
9. **Kauzmann, W.,** Some factors in the interpretation of protein denaturation, *Adv. Protein Chem.,* 14, 1, 1959.
10. **Flory, P. J.,** Thermodynamics of high polymer solutions, *J. Chem. Phys.,* 10, 51, 1942.
11. **Huggins, M. L.,** Some properties of solutions of long-chain compounds, *J. Phys. Chem.,* 46, 151, 1942.
12. **Hamaker, H. C.,** The London-van der Waals attraction between spherical particles, *Physica,* 4, 1058, 1937.
13. **Reichardt, C.,** in *Solvent Effects in Organic Chemistry,* Reichardt, C., Ed., Vol. 3, Verlag Chemie, New York, 1979.
14. **Meyer, K. H.,** Über das Gleichgewicht desmotroper Verbindungen in verschiedenen Lösungsmitteln, *Ber. Dtsch. Chem. Ges.,* 47, 826, 1914.
15. **Eliel, E. L. and Hofer, O.,** Conformational analysis. XXVII. Solvent effects in conformational equilibria of heterosubstituted 1, 3-dioxanes, *J. Am. Chem. Soc.,* 95, 8041, 1973.
16. **Gutmann, V.,** in *Coordination Chemistry in Non-Aqueous Solutions,* Springer-Verlag, New York, 1968.
17. **Llor, J. and Cortijo, M.,** A model process for measurement of solvent polarity, *J. Chem. Soc., Perkin Trans.,* 2, 1111, 1977.
18. **Winstein, S. and Fainberg, A. H.,** Correlation of solvolysis rates. IV. Solvent effects on enthalpy and entropy of activation for solvolysis of *t*-butyl chloride, *J. Am. Chem. Soc.,* 79, 5937, 1957.
19. **Grunwald, E. and Winstein, S.,** The correlation of solvolysis rates, *J. Am. Chem. Soc.,* 70, 846, 1948.
20. **Winstein, S., Grunwald, E., and Jones, H. W.,** The correlation of solvolysis rates and the classification of solvolysis reactions into mechanistic categories, *J. Am. Chem. Soc.,* 73, 2700, 1951.
21. **Drougard, Y. and Decroocq, D.,** L'influence du solvant sur la reaction chimique. II. Etude et correlation des effets physiques du milieu, *Bull. Soc. Chim. France,* 1969, 2972.
22. **Gielen, M. and Nasielski, J.,** Electrophilic substitution at a saturated carbon atom. III. Competition between steric and inductive effects as a function of the polarity of solvents in the reaction of halogens with tetraalkyltin compounds, *J. Organometal. Chem.,* 1, 173, 1963.
23. **Berson, J. A., Hamlet, Z., and Mueller, W. A.,** The correlation of solvent effects on the stereoselectivities of Diels-Alder reactions by means of linear free energy relationship. A new empirical measure of solvent polarity, *J. Am. Chem. Soc.,* 84, 297, 1962.
24. **Kosower, E. M.,** The effect of solvent on spectra. I. A new empirical measure of solvent polarity: Z-values, *J. Am. Chem. Soc.,* 80, 3253, 1958.
25. **Mohammad, M. and Kosower, E. M.,** Stable free radicals. VI. The reaction between 1-ethyl-4-carboe-thoxypyridinyl radical and 4-nitrobenzyl halides, *J. Am. Chem. Soc.,* 93: 2713, 1971.
26. **Dimroth, K., Reichardt, C., Siepmann, T., and Bohlmann, F.,** Über Pyridinium-N-Phenyl-Betaine und ihre Verwendung zur Charakterisierung der Polarität von Lösungsmitteln, *Liebigs Ann. Chem.,* 661, 1, 1963.
27. **Kamlet, M., Abboud, J. L., and Taft, R. W.,** The Solvatochromic comparison method. 6. The π^* scale of solvent polarities, *J. Am. Chem. Soc.,* 99, 6027, 1977.
28. **Kovats, E.,** Gas chromatographic characterization of organic substances in the relation index system, *Adv. Chromatog.,* 1, 229, 1965.
29. **Kovats, E. and Weisz, P. B.,** Über den Retentionsindex und seine Verwendung zur Ausstellung einer Polaritätsskala für Lösungsmittel, *Ber. Bunsenges. Phys. Chem.,* 69, 812, 1965.
30. **Rohrschneider, L.,** Solvent characterization by gas-liquid partition coefficients of selected solutes, *Anal. Chem.,* 45, 1241, 1973.
31. **Snyder, L. R.,** Classification of the solvent properties of common liquids, *J. Chromatog. Sci.,* 16, 223, 1978.
32. **Abboud, J. L., Kamlet, M. J., and Taft, R. W.,** Regarding a generalized scale of solvent polarities, *J. Am. Chem. Soc.,* 99, 8325, 1977.
33. **Melander, W. and Horvath, C.,** Salt effects on hydrophobic interactions in precipitation and chromatography of proteins: an interpretation of the lyotropic series, *Arch. Biochem. Biophys.,* 183, 200, 1977.

34. **Lewin, S.**, in *Displacement of Water and its Control of Biochemical Interactions*, Lewin, S., Ed., Academic Press, New York, 1974.

35. **Srinivasan, R. and Ruckenstein, E.**, Role of physical forces in hydrophobic interaction chromatography, *Separat. Purif. Methods*, 9, 67, 1980.

36. **Hatefi, Y. and Hanstein, W. G.**, Solubilization of particulate proteins and nonelectrolytes by chaotropic agents, *Proc. Nat. Acad. Sci.*, 62, 1129, 1969.

37. **Damodaran, S. and Kinsella, J. E.**, Effects of ions on protein conformation and functionality, in *Food Protein Deterioration*, Cherry, J. P., Ed., American Chemical Society, Washington, D.C., 1982, 13.

38. **Furukawa, T. and Ohta, S.**, Solubility of isolated soy protein in ionic environments and an approach to improve its profile, *Agric. Biol. Chem.*, 47, 751, 1983.

39. **Kinsella, J. E. and Hale, M. L.**, Hydrophobic associations and gluten consistency: effects of specific anions, *J. Agric. Food Chem.*, 32, 1054, 1984.

40. **Arntfield, S. D., Murray, E. D., and Ismond, M. A. H.**, Effect of salt on the thermal stability of storage proteins from fababean *(Vicia Faba)*, *J. Food Sci.*, 51, 371, 1986.

41. **Wetlaufer, D. B., Malik, S. K., Stoller, L., and Coffin, R. L.**, Nonpolar group participation in the denaturation of protein by urea and guanidinium salts. Model compound studies, *J. Am. Chem. Soc.*, 86, 508, 1964.

42. **Frank, H. S. and Franks, F.**, Structural approach to the solvent power of water for hydrocarbons; urea as a structure breaker, *J. Chem. Phys.*, 48, 4746, 1968.

43. **Robinson, D. R. and Jencks, W. P.**, The effect of compounds of the urea-guanidinium class on the activity coefficient of acetyltetraglycine ethyl ester and related compounds, *J. Am. Chem. Soc.*, 87, 2462, 1965.

44. **Roseman, M. and Jencks, W. P.**, Interactions of urea and other polar compounds in water, *J. Am. Chem. Soc.*, 97, 631, 1975.

45. **Kuharski, R. A. and Rossky, P. J.**, Solvation of hydrophobic species in aqueous urea solution: a molecular dynamics study, *J. Am. Chem. Soc.*, 106, 5794, 1984.

Chapter 2

QUANTITATIVE ESTIMATION OF HYDROPHOBICITY

TABLE OF CONTENTS

I. HYDROPHOBICITY SCALES FOR AMINO ACID RESIDUES

Hydrophobicity scales that have been published can be classified into two groups: (1) those derived from solubility measurement and (2) those calculated empirically from the molecular structure, especially X-ray patterns, based on the assumption that hydrophobic amino acid residues are buried in the interior of protein molecules. Since this assumption has not been totally confirmed, the empirical scales are not representing, in the strict sense, the definition of hydrophobicity. However, it is customary to deduce that the hydrophobic residues are buried when the two parameters, the measured and empirical scale values, are closely correlated.

Whenever there are deviations between these two parameters, it is concluded that some hydrophobic residues can locate on the surface of protein molecules, or some hydrophilic residues may be buried inside of the molecules. It may be safe to state that the polarity of amino acid residues and the extent of buriedness are not completely parallel, therefore the above two scales may not necessarily be interpreted in a similar manner. It may also be important to bear in mind that the "buriedness" is difficult to define and, therefore, the second type of hydrophobicity scales are more empirical than the first type which are measured scales.

Rose et al.[1] compared 15 different scales, 3 measured, 8 empirical, and 4 measured/empirical. Tables 1, 2, and 3 were refitted from their tables with the additional data of Rose et al.[17] It is extremely difficult to find any trend in these tables. Even within measured or empirical values, great differences are observed; e.g., the correlation coefficient of only −0.04 is found between the scales of Nozaki-Tanford, and Wolfenden et al., both of which are measured scales. Nozaki and Tanford[2] found tryptophan and tyrosine to be very hydrophobic, while Wolfenden et al.[13] reported these residues to be very hydrophilic.

To overcome this inconsistency, various attempts to compromise the situation have been made:

1. One such attempt has been to use a new definition of the buriedness to secure a better correlation with a measured scale. A recent example is the paper by Rose et al.[17] The plot of the average area buried upon folding as a function of the Nozaki-Tanford scale for 11 hydrophobic residues showed a good correlation.
2. Sweet and Eisenberg[16] used a stepwise replacement method to compute an "optimal matching hydrophobicity". In Table 3, this scale shows the smallest standard deviation of 0.182, implying good correlations with all other scales. The scale of Meirovitch et al.[11] shows the best average correlation coefficient of 0.72 with a reasonable standard deviation of 0.188 and
3. Another compromise is to use a combination of several scales separately. For computation of hydrophobicity profile for prediction of turns, Rose et al.[1] used Nozaki-Tanford, Hopp-Woods and Kyte-Doolittle scales. The reasons for this selection were not clearly stated.

Unfortunately, there is no standard rule for the selection of a hydrophobicity scale. However, it is observed in Table 3 that the correlation coefficients between scales 1 to 10 are relatively high, thus making a group. Similarly, scales 11 to 14 make another group with high correlation coefficients between scales. However, scales 15 and 16 do not belong to any group, nor do they have a good correlation between themselves.

II. INTRINSIC FLUORESCENCE OF PROTEINS

Due to deficiency of electrons in the molecules, unsaturated compounds absorb radiation energy. Compared to single double-bond, conjugated double-bonds require less energy, thus

Table 1
SCALES OF HYDROPHOBICITY FOR THE AMINO ACID RESIDUES

Type of scale[a]	Scale parameter	Number of residues in scale	Ref.
Measured	ΔG (transfer from H_2O) to organic solvent	11	Nozaki and Tanford[2]
Measured/empirical	Extended scale of Nozaki and Tanford[2]	20	Jones[3]
Empirical	Fraction of residues at least 95% buried[b]	20	Chothia[4]
Measured/empirical	Combination of Nozaki and Tanford[2] and Chothia[4]	20	Levitt[5]
Measured/empirical	Average of scale values from Nozaki and Tanford[2] for residues with an 8 Å sphere	20	Manavalan and Ponnuswamy[6]
Empirical	Fraction of residues that are buried	20	Wertz and Scheraga[7]
Empirical	$\Delta G_{stabilization}$ from pairwise contacts	20	Krigbaum and Komoriya[8]
Empirical	$\Delta G_{transfer}$ from buried interior to solvent-accessible surface[b]	20	Janin[9]
Empirical	Information theoretic measure of distribution between interior and surface	20	Robson and Osguthorpe[10]
Empirical	Reduced distance from center of mass and average side-chain orientation angle	20	Meirovitch et al.[11]
Empirical	Average contact number of α-carbons within an 8 Å sphere	20	Nishikawa and Ooi[12]
Measured	$\Delta G_{transfer}$ from H_2O to vapor	19	Wolfenden et al.[13]
Measured	$\Delta G_{transfer}$ from H_2O to octanol	16	Yunger and Cramer[14]
Measured/empirical	Modified scale of Wolfenden et al.[13]	20	Kyte and Doolittle[15]
Empirical	Average of hydrophobicity values for residue replacements from related structures	20	Sweet and Eisenberg[16]
Empirical	Average area that each residue buries upon folding	20	Rose et al.[17]

[a] Each scale is classified as a solution measurement, an empirical calculation, or a combination of the two.
[b] Residue accessibility is measured using the method of Lee and Richards.[18]

absorb the radiation energy at longer wavelengths, e.g., 217 nm for butadiene ($H_2C = CH$ $- CH = CH_2$) compared to 195 nm for ethylene ($CH_2 = CH_2$). The benzenoid form of benzene absorbs at an even longer wavelength of 255 nm. Among all of the amino acids composing protein molecules, aromatic amino acids are the only unsaturated amino acids responsible for UV absorption. Substitution on the same radical also increases the wavelength of maximum absorption, thus phenylalanine, tyrosine, and tryptophan absorb radiation energy at longer wavelengths in this order (Figure 1). Since fluorescence is caused by emission of radiation from the lowest vibrational state after giving energy from the excited state to the surrounding solvent molecules, the emission wavelength is longer than the excitation wavelength (Figure 1).

The major contribution to the intrinsic fluorescence of proteins comes from tryptophan residues, which arises from the lowest $\pi - \pi^*$ transition of the indole side chain. The location of tryptophan residues in protein molecules affects the fluorescence energy.[19] Three spectral classes were reported, one buried in nonpolar regions of the protein ($\lambda_{em} = 330$—332 nm), one completely exposed to surrounding water ($\lambda_{em} = 350$—353 nm) and another in limited contact with water which is probably immobilized at the surface of the protein molecule ($\lambda_{em} = 340$—342 nm).

Table 2
HYDROPHOBICITY SCALE VALUES OF AMINO ACID RESIDUES

	$\Delta G'_i$ (3)	$\Delta G'_i$ (2)	ξ_i (8)	Fraction buried (7)	$<r_f>$ (11)	$\Delta G'_i$ (10)	$\Delta G'_i$ (5)	H kcal (6)	OMH (16)	A°−<A> (17)	Fraction 95% buried (4)	$\Delta G'_i$ (9)	Hydropathy Index (15)	$\Delta G'_i$ (13)	log P (14)	single residue parameter (12)
Ala	0.87	0.5	4.32	0.52	0.93	-1.0	-0.5	12.97	-0.40	86.6	0.38	0.59	1.8	1.12	-2.74	0.23
Arg	0.85	—	6.55	0.49	0.98	0.3	3.0	11.72	-0.59	162.2	0.01	-1.82	-4.5	-2.55	-1.99	-0.26
Asn	0.09	—	6.24	0.42	0.98	-0.7	0.2	11.42	-0.92	103.3	0.12	-0.55	-3.5	-0.83	—	-0.94
Asp	0.66	—	6.04	0.37	1.01	-1.2	2.5	10.85	-1.31	97.8	0.15	-0.69	-3.5	-0.83	—	-1.13
Cys	1.52	—	1.73	0.83	0.88	2.1	-1.0	14.63	0.17	132.3	0.45ᵃ	1.44	2.5	0.59	—	1.78
Gln	0.00	—	6.13	0.35	1.02	-0.1	0.2	11.76	-0.91	119.2	0.07	-0.83	-3.5	-0.78	—	-0.57
Glu	0.67	—	6.17	0.38	1.02	-0.7	2.5	11.89	-1.22	113.9	0.18	-0.83	-3.5	-0.92	-0.94	-0.75
Gly	0.10	0.0	6.09	0.41	1.00	0.0	0.0	12.43	-0.67	62.9	0.36	0.59	-0.4	1.20	-3.11	-0.07
His	0.87	0.5	5.66	0.70	0.89	1.1	-0.5	12.16	-0.64	155.8	0.17	0.02	-3.2	-0.93	-1.91	0.11
Ile	3.15	—	2.31	0.79	0.79	4.0	-1.8	15.67	1.25	158.0	0.60	1.16	4.5	1.16	-1.69	1.19
Leu	2.17	1.8	3.93	0.77	0.85	2.0	-1.8	14.90	1.22	164.1	0.45	0.87	3.8	1.18	-1.79	1.03
Lys	1.64	—	7.92	0.31	1.05	-0.9	3.0	11.36	-0.67	115.5	0.03	-2.39	-3.9	-0.80	-1.15	-1.05
Met	1.67	1.3	2.44	0.76	0.84	1.8	-1.3	14.39	1.02	172.9	0.40	0.73	1.9	0.55	-1.87	0.66
Phe	2.87	2.5	2.59	0.87	0.78	2.8	-2.5	14.00	1.92	194.1	0.50	0.87	2.8	0.67	-1.43	0.48
Pro	2.77	—	7.19	0.35	1.00	0.4	-1.4	11.37	-0.49	92.9	0.18	-0.26	-1.6	0.54	-2.54	-0.76
Ser	0.07	-0.3	5.37	0.49	1.02	-1.2	0.3	11.23	-0.55	85.6	0.22	0.02	-0.8	-0.05	-3.07	-0.67
Thr	0.07	0.4	5.16	0.38	0.99	-0.5	-0.4	11.69	-0.28	106.5	0.23	-0.12	-0.7	-0.02	-2.94	-0.36
Trp	3.77	3.4	2.78	0.86	0.83	3.0	-3.4	13.93	0.50	224.6	0.27	0.59	-0.9	-0.19	-1.11	0.90
Tyr	2.67	2.3	3.58	0.64	0.93	2.1	-2.3	13.42	1.67	177.7	0.15	-0.40	-1.3	-0.23	-2.03	0.59
Val	1.87	1.5	3.31	0.72	0.81	1.4	-1.5	15.71	0.91	141.0	0.54	1.02	4.2	1.13	-2.26	1.24

Note: Reference numbers are in parentheses.

ᵃ 0.50 and 0.40 for Cys and 1/2 Cys, respectively.

Table 3
CORRELATION MATRIX[a] OF HYDROPHOBICITY SCALES FOR AMINO ACID RESIDUES

Scale	Type[b]	(1)	(2)	(3)	(4)	(5)	(6)	(7)	(8)	(9)	(10)	(11)	(12)	(13)	(14)	(15)	(16)	Mean ± SD
(1) Jones[3]	M/E		.99	.53	.66	.68	.83	.70	.63	.75	.72	.41	.35	.47	.34	.47	.40	.60 ± .190
(2) Nozaki and Tanford[2]	M			.81	.82	.75	.91	.99	.66	.78	.91	.20	.28	.24	.04	.89	.63	.66 ± .134
(3) Krigbaum and Komoriya[8]	E				.89	.87	.78	.74	.86	.78	.61	.79	.81	.79	.63	.15	.04	.67 ± .255
(4) Wertz and Sheraga[7]	E					.93	.88	.72	.85	.80	.78	.71	.74	.71	.49	.35	.15	.70 ± .215
(5) Meirovitch et al.[11]	E						.87	.75	.89	.82	.73	.79	.75	.78	.56	.29	.34	.72 ± .188
(6) Robson and Osguthorpe[10]	E							.74	.83	.84	.80	.65	.64	.64	.48	.40	.21	.70 ± .201
(7) Levitt[5]	M/E								.67	.75	.52	.62	.77	.67	.65	.05	.14	.63 ± .240
(8) Manavalan and Ponnuswamy[6]	M/E									.82	.58	.86	.76	.71	.25	.23	.82	.70 ± .208
(9) Sweet and Eisenberg[16]	E										.71	.67	.58	.75	.57	.22	.34	.68 ± .182
(10) Rose et al.[17]	E											.24	.23	.27	.06	.66	.58	.56 ± .249
(11) Chothia[4]	E												.91	.96	.86	.06	.00	.58 ± .319
(12) Jannin[9]	E													.87	.83	.22	.15	.59 ± .269
(13) Kyte and Doolittle[15]	M/E														.89	.12	.07	.60 ± .290
(14) Wolfenden et al.[13]	M															.30	.10	.47 ± .283
(15) Yunger and Cramer[14]	M																.43	.32 ± .227
(16) Nishikawa and Ooi[12]	E																	.29 ± .241

[a]The absolute values of correlation coefficient are shown.
[b]M: solubility measurement, E: empirical calculation.

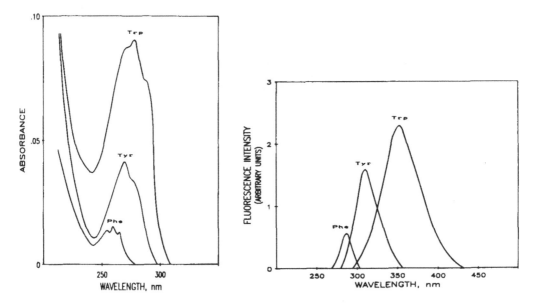

FIGURE 1.   Absorption and emission spectra (fluorescence intensity) of Phe, Tyr, and Trp residues of a peptide. Adapted from Genot et al.[24]

In general, the intensity of tryptophan fluorescence and the wavelength of maximum intensity are highly dependent on the polarity of the environment. Apolipoprotein-alanine from human very-low-density lipoprotein with three tryptophan residues shows the wavelength of maximum emission at 343.5 nm. In the presence of phosphatidylcholine this maximum is blue-shifted to 335 nm and the magnitude of the intensity increased by 2.8-fold.[20] This principle has been utilized to study lipid-protein interactions, e.g., membrane protein[21] and wheat gluten.[22]

III. FLUORESCENCE QUENCHING

The presence of groups that tend to withdraw electrons, e.g., carboxyl, azo and nitro groups and halides, usually quenches fluorescence. Influence of neutral salt solutions on the intrinsic fluorescence of proteins was investigated.[23] The results with tryptophan and tyrosine indicated the involvement of a collisional quenching mechanism due to agreement with the Stern-Volmer law:

$$F_o/F = 1 + k_q(X) \tag{1}$$

where F_o and F are quantum yields of fluorescence in the absence and presence of quencher (X), respectively, and k_q is the quenching constant. The effectiveness of anions in quenching fluorescence of proteins followed similar sequences which almost resembled the Hofmeister series. Influence of monovalent cations was indicative of electrostatic and lyotropic effects of ions. However, ionic quenchers, being charged and heavily hydrated, quench only surface tryptophanyl residues and the quenching action may be influenced by electrostatic effects of proteins. To circumvent this problem, uncharged quenchers, e.g., acrylamide[24] and 2,2,2-trichloroethanol,[25] were used to determine the exposure of tryptophan residues in protein molecules. It is, however, unlikely that very useful quantitative information will be obtained on the overall protein hydrophobicity by using this method.

IV. HYDROPHOBIC PROBE METHODS

Hydrophobic probe methods may be the simplest type of method for measuring protein hydrophobicity. Among many hydrophobic probes which have been used so far, 1-anilinonaphthalene-8-sulfonate (ANS) is the most frequently used probe. The method is based on the rule that the fluorescence emission of ANS and its analogues depends critically on environment. Whereas, the quantum yields of fluorescence can exceed 0.70 in a nonpolar solvent, the emissions are strongly quenched in water to give yields in the range of 0.0032 to 0.011.[26] The emission maxima are also shifted as much as 100 nm to longer wavelength (red shift) in water.

Recent examples of the application of this method in protein chemistry are the monitoring of heat denaturation of serum albumin,[27] the measurement of exposed hydrophobic regions in histone oligomers,[28] and the determination of the effects of iodination on thyroglobulin conformation.[29] Heat denaturation decreased the affinity of serum albumin for ANS and this was interpreted as a result of loss of the hydrophobic region and the loss of dye-binding sites due to aggregation through hydrophobic interaction and intermolecular disulfide bond formation. In the case of histone oligomers, the difference in the enhanced ANS fluorescence was attributed to the difference in exposed hydrophobic regions in the molecules. It was observed in thyroglobulin that decreasing the iodine content increased the number of ANS-binding sites. This phenomenon was discussed in relation to hydrophobic forces.

However, direct linking of ANS results with hydrophobicity may need great caution, as Penzer[30] warned. He reported that the fluorescence emission was enhanced and blue-shifted in strong aqueous $MgCl_2$ solutions which was similar to the case when the solvent was changed from water to nonpolar solvents. ANS in solutions has two different conformations. The most unhindered shape is the one in which the planes of the benzene and naphthalene rings are approximately perpendicular (I). The other shape is one in which the two rings are nearly coplanar (II) because of hydrogen bonding between NH and SO_3 groups. Solvents which favor intermolecular hydrogen bond formation will favor formation of the more rigid conformation (II) which is fluorescent.

$$I \qquad\qquad II \qquad (2)$$

Therefore, it was argued that molecular rigidity rather than solvent polarity is the dominant factor influencing the energy and the quantum yield of fluorescence of ANS.

The use of cis-parinarate (CPA) as a hydrophobic probe was suggested by Sklar et al.[31]

$$
\begin{array}{c}
 \text{H} \;\; \text{H} \;\; \text{H} \;\;\;\; \text{H} \\
 | \;\;\; | \;\;\; | \;\;\;\;\; | \\
\text{H}_3\text{C} \cdot \text{CH}_2 \cdot \text{C=C--C=C--C=C--C=C} \cdot (\text{CH}_2)_7 \cdot \text{COOH} \\
 | \;\;\;\;\;\; | \;\; | \;\; | \\
 \text{H} \;\;\;\;\; \text{H} \; \text{H} \; \text{H}
\end{array}
\tag{3}
$$

9,11,13,15-*cis, trans, trans, cis*-octadecatetraenoic acid (*cis*-Parinaric acid).

Since the structure of the parinaric acid isomers are very similar to naturally occurring fatty acids, the perturbing effects of these molecules on biological membranes should be minimal. This fact was substantiated by Tecoma et al.,[32] who demonstrated that parinaric acid could be incorporated directly into the cellular phospholipids of an *Escherichia coli* fatty acid auxotroph. Phase transitions detected in *E. coli* membranes containing the esterified parinaroyl-phospholipid probe were identical to those determined with the free fatty acid probe. The usefulness of CPA as a probe of membrane fluidity was established[33] by comparison of its fluorescent properties in phosphatidylcholine vesicles with those of the more commonly used fluorescent probe, 1,6-diphenyl-1,3,5-hexatriene (DPH).[34]

Other hydrophobic probes for specific uses are 2-*p*-toluidylnaphthalene-6-sulfonate (TNS)[35] and 5,5'-*bis*[8-(phenylamino)-1-naphthalene-sulfonate] [*bis*(ANS)] which is less dissociable than ANS.[36]

V. HYDROPHOBIC CHROMATOGRAPHY

In partition chromatography, partition coefficient (or distribution constant) is the principal contributing factor for chromatographic separation. Figure 2 illustrates an effect of the partition coefficient on the resolution of chromatography. A computer program was written for a hypothetical column composed of 10 cells; within each cell, the distribution equilibrium was achieved between the mobile and stationary phases before the mobile phase was shifted to the following cell for establishing the new equilibrium. The three elution patterns represent the cases when solute A had partition coefficients K_A ($= c_s/c_m$, where c_s and c_m are the concentrations of the solute in the stationary and mobile phases, respectively) of 0.67, 0.3, and 0.15, while the partition coefficient of solute B (K_B) was constant at 1.5. As seen in Figure 2, the larger the difference between K_A and K_B, the better the resolution. Although the number of plates (N) is another important factor for good resolution, it improves the sharpness of each peak, but not the elution time (or volume) difference. N in the case of this model example is equivalent to the number of cells. Partition coefficient is measured as a solubility ratio between water (polar) and an immiscible organic solvent (nonpolar).

The partition coefficient can be written also as

$$
P = C_s/C_w
\tag{4}
$$

where C_s and C_w are the equilibrium concentrations of the solute in the organic and aqueous phases, respectively. A solute with high P is regarded as lipophilic (or hydrophobic) and a solute with low P as hydrophilic. As the P scale covers a range of more than 10^{10}, logarithmic P values are usually used. Since the partition coefficient is an equilibrium constant (K) and $\Delta G° = -RT \ln K$, log P is proportional to the free energy of transfer from water phase to organic solvent phase ($\Delta G°$). The free energy of transfer can be split to the fragments:

$$
\log P = \Sigma a_n f_n
\tag{5}
$$

where f_n is the hydrophobic fragmental constant of n fragments, the lipophilicity contribution of a constituent part of the structure of compounds to the total lipophilicity, and "a" is a

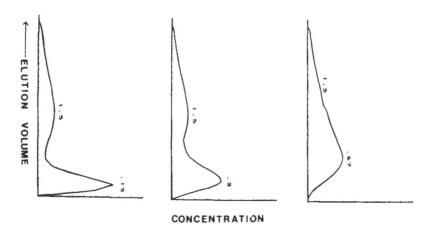

FIGURE 2. Change in resolution of two peaks due to the change in partition coefficient difference. The patterns were computed for elution from a hypothetical column composed of 10 cells after complete partition equilibration in each cell.

numerical factor indicating the incidence of a given fragment in the structure. An example of the computation using Equation 5 is shown in Equation 7 for a compound with the structure:

$$\underset{C_6H_5}{\overset{C_6H_5}{\diagdown}}CH-O-CH_2-CH_2-N\underset{CH_3}{\overset{CH_3}{\diagup}} \quad (6)$$

$$\log P = 2f(C_6H_5) + f(CH) + f(O) + 2f(CH_2) + f(N) + 2f(CH_3) + p.e.2$$

$$= 3.792 + 0.236 + (-1.536) + 1.054 + (-2.133) + 1.404 + 0.46$$

$$= 3.28 \quad (7)$$

This calculated value is in excellent agreement with the experimental values of 3.27 and 3.40; p·e·2 is the proximity correction for a 2C separation.[37] High correlations between log P calculated by the method of Rekker[37] and log k' ($k' = (t_R - t_O)/t_O$, where t_R and t_O are elution times of retained and unretained peaks, respectively) were reported for alkylbenzyne derivatives, implicating the validity of the theory.[38]

Since the success of attaching alkyl groups of different length and aryl groups to column material, e.g., gel and porous silica,[39,40] reverse phase chromatography has become a useful tool for analysis and separation of compounds. To elute solutes, polarity of the eluents is decreased by increasing the proportion of miscible nonpolar solvent in the aqueous eluent. However, in the case of protein separation, an ionic strength change by decreasing salt concentration in the eluents is also used for eluting the solutes to avoid denaturation of proteins with organic solvents. This technique is called "hydrophobic interaction chromatography" (HIC) to distinguish it from the regular reverse phase chromatography (RPC). Figure 3 illustrates the difference in the principle among normal phase chromatography, RPC and HIC.

As expected, the elution behavior of peptides by using reverse phase chromatography showed good agreement with the predicted hydrophobicity values computed for constituent amino acid residues.[41-45] Figure 4 is an example of the peptide retention time reported by Meek.[43]

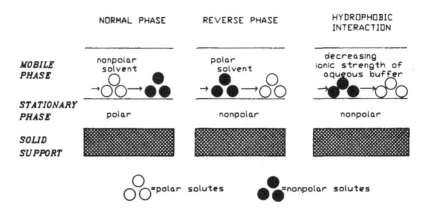

FIGURE 3.   Comparison of the separation principle among normal phase, reverse phase, and hydrophobic interaction chromatography.

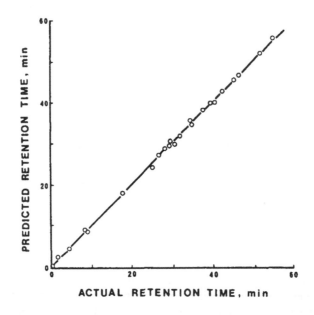

FIGURE 4.   Relationship between actual retention times of peptides and times predicted by summing retention coefficients of the amino acids and end groups (pH 7.4). Adapted from Meek.[43]

Wilson et al.[45] investigated the behavior of 96 peptides, ranging in length from 2 to 65 amino acid residues. After computing hydrophobic constants of amino acid residues from the retention properties of these peptides, they were compared with other hydrophobic constants published in the literature (Table 4). Correlation coefficients in the upper row show that the values of Wilson et al.,[45] calculated by multiple regression analysis, best explain the elution behavior of the peptides. Correlation coefficients of the hydrophobicity constants for amino acid residues determined by Wilson et al.[45] was computed vs. other constants and shown in the bottom row. Correlation coefficients of 0.756, 0.776, and 0.674 were also obtained vs. the values of Jones,[3] Meirovitch et al.,[13] and Kyte and Doolittle,[15] respectively. The lower value for Kyte-Doolittle is due to their lower assignment of hydrophobic scores for tryptophan and tyrosine.

Summarizing all of the above data, it may be justified to conclude that reverse phase

Table 4
AMINO ACID HYDROPHOBICITY CONSTANTS IN RELATION TO PEPTIDE
CHROMATOGRAPHIC RETENTION TIMES

	Bigelow-Chapman[46]	Meek[43]	Pliska-Fauchere[47]	Rekker[37]	Segrest-Feldman[48]	Wilson et al.[45]
Ala	0.5	−0.1	0.38	0.53	1.0	−0.3
Arg	0.75	−4.5	−1.23	−0.82	—	−1.1
Asn	—	−1.6	−0.27	−1.05	−1.5	−0.2
Asp	0	−2.8	−1.23	−0.02	—	−1.4
Cys	—	−2.2	—	1.11	0	6.3
Gln	—	−2.5	−0.09	−1.09	−1.0	−0.2
Glu	0	−7.5	−1.20	−0.07	—	0
Gly	0	−0.5	0	0	0	1.2
His	0.50	0.8	−1.3	−0.23	1.0	−1.3
Ile	2.95	11.8	1.56	1.99	5.0	4.3
Leu	1.80	10.0	1.66	1.99	3.5	6.6
Lys	1.50	−3.2	−0.93	−0.52	—	−3.6
Met	1.30	7.1	1.39	1.08	2.5	2.5
Phe	2.50	13.9	1.80	2.24	5.0	7.5
Pro	2.60	8.0	0.56	1.01	1.5	2.2
Ser	−0.30	−3.7	0.04	−0.56	−0.5	−0.6
Thr	0.40	1.5	−0.33	−0.26	0.5	−2.2
Trp	3.40	18.1	1.87	2.31	6.5	7.9
Tyr	2.30	8.2	1.70	1.70	4.5	7.1
Val	1.50	3.3	1.06	1.46	3.0	5.9
Correlation coefficient						
With retention time	0.536	0.681	0.713	0.693	0.826	0.831
With Wilson et al.[45]	0.741	0.757	0.868	0.841	0.778	—

chromatographic behavior of peptides can be explained from the polarity of the constituent amino acid residues and that the hydrophobic scales of Nozaki-Tanford and its analogues are reasonable for evaluating protein hydrophobicity, in a similar sense.

Chromatographic behavior of proteins is not as straightforward as it is for small peptides. When the retention times of 12 proteins were compared between RPC with isopropanol-water as an eluent and HIC,[49] there was no good correlation ($r = 0.24$). Recovery of enzyme activity was greater than 86% for lactic dehydrogenase and α-chymotrypsin after HIC, while none of the activity of β-glucosidase was recovered after RPC. Similarly, Luiken et al.[50] reported that although the activity of ribonuclease was unaffected, only 1 to 5% of the activity of horseradish peroxidase was recovered after RPC. Sadler et al.[51] reported that propanol, the most popular solvent for RPC of proteins because of their easy dissolution, induced a reversible conformational change in proteins to an apparently ordered, helical form. Even in the case of HIC that was supposed to be milder in the elution conditions than RPC, the hydrophobic column itself, rather than temperature, caused unfolding of some proteins.[52] Although successful to show the hydrophobicity differences between proteins, it is still not clear whether the hydrophobic parameter derived from the elution time during HIC is actually related to their native hydrophobicity.[53] In addition, difference in binding between aliphatic and aromatic adsorbents was observed.[54] Although aromatic (π-π) effect, per se, is presumably a direct electronic interaction and not a lyotropic effect, it is possible that π-π interaction, and hydrophobic interaction of the phenyl group, reinforce each other during chromatographic separation.

VI. HYDROPHOBIC BINDING METHODS

A. Hydrocarbons and Their Analogs

Binding of aroma compounds with food components, especially macromolecular compounds, was reported.[55] Sorption of ethanol, acetone, hexane, and ethyl acetate depended on the hydrophobicity of proteins;[56,57] The hydrophobicity scale used was that of Bigelow.[58]

Mohammadzadeh-K. et al.[59,60] compared a variety of aliphatic and aromatic hydrocarbons in terms of binding with BSA and β-lactoglobulin, and n-heptane was chosen to measure the binding capacity of proteins. Marked influence of protein structure on the hydrocarbon binding capacity was observed.

B. Sodium Dodecylsulfate

Binding of sodium dodecylsulfate (SDS) with proteins is primarily hydrophobic in nature.[61,62] SDS binding capacity showed a good correlation with the surface hydrophobicity measured fluorometrically with *cis*-parinarate.[63] An advantage of the SDS binding method is the capability of applying it to insoluble proteins.

C. Triglycerides

Binding capacity of proteins for triglycerides was measured by Smith et al.[64] The method entails sonication of an aqueous suspension of a triglyceride and equilibrating the microemulsion with a protein solution. After filtering through a polycarbonate membrane filter, the triglyceride stabilized with protein in the filtrate is analyzed either by GLC or by a radioactive technique. It was found that an increase in molecular weight of saturated triglycerides or an increase in double bonds of unsaturated triglycerides reduced hydrophobic binding. Trioctanoin was selected for binding study of a series of proteins.

Binding capacity of proteins for corn oil was determined by using a nonpolar, nondissociable fluorescent probe 1,6-diphenyl-1,3,5-hexatriene (DPH) as a quantitative marker of bound oil.[65] The hydrophobicity of proteins measured by this method showed a better correlation with ANS hydrophobicity than with CPA hydrophobicity. An advantage of this method is the simplicity in measurement.

VII. HYDROPHOBIC PARTITION

As discussed in Section V, partition coefficient determination may be a preferable method for determining hydrophobicity of proteins. However, the need for use of water-immiscible organic solvents may cause a problem, since most proteins do not dissolve in the organic solvents. Selection of the water-immiscible phase is critical; addition of hydrophobic polymers to aqueous solutions can cause a phase separation. Polyethylene glycol (PEG), dextran, and Ficol (a synthetic polymer of sucrose) can form a three-phase PEG-dextran-Ficol system from the top to the bottom after centrifugation.

Shanbhag and Axelsson[66] used a pair of two-phase systems containing dextran and PEG with and without esterified fatty acids for determination of hydrophobic effects of proteins. The principle of this method is schematically illustrated in Figure 5. $\Delta\log K$ is computed as an index of hydrophobicity of protein.

$$\Delta\log K = \log K'' - \log K'$$
$$= \log(c_u/c_l)'' - \log(c_u/c_l)' \qquad (8)$$

where c_u and c_l are concentrations of the protein at equilibrium in the upper and lower phase, respectively.

Similarly, Zaslavsky et al.[67] used an aqueous two-phase system of dextran and Ficol.

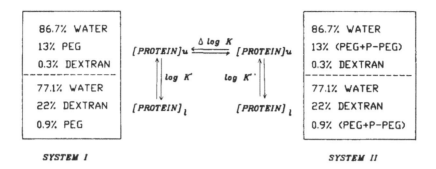

FIGURE 5. Schematic presentation of the determination of hydrophobic interaction. The total composition of system I is 8% dextran T70, 8% poly(ethylene glycol) 6000, and 84% water. In system II a part of the poly(ethylene glycol) has been replaced by poly(ethylene glycol) palmitate, the rest of the system being identical. The composition of the upper and lower phases in the respective system are indicated in the figure. $\Delta G'$ and $\Delta G''$ are the free energy of transfer for a protein from the lower phase to the upper phase in system I and II respectively; PEG = poly(ethylene glycol); P-PEG = poly(ethylene glycol) palmitate. Adapted from Shanbhag and Axelsson.[66]

VIII. COMPARISON OF THE EMPIRICAL METHODS

Intrinsic fluorescence and fluorescence quenching methods both measure the behavior of aromatic amino acid residues in protein molecules, and thus are not directly relating to the entire hydrophobicity of the proteins.

Hydrophobic probe methods using ANS, CPA, or similar probes are simple in the procedure. Furthermore, because of a long history of utilization in protein studies, the ANS method may be the most popular and widely used method. The use of CPA may be advantageous for hydrophobicity measurement of food proteins due to its structural similarity to naturally occurring fatty acids and the consequent possible analogy to protein-fatty acid and protein-lipid interactions in food systems. However, the frequent inconsistency between the values of ANS hydrophobicity and CPA hydrophobicity for some proteins,[65,68] requires reasonable explanation. In addition, the possible contribution of the charged moieties to the interaction between these anionic probes with proteins must be noted.

Advantages and disadvantages of chromatographic scales for protein hydrophobicity have already been discussed. The major criticism is the possible denaturation of protein molecules during chromatography. Although hydrophobic interaction chromatography (HIC) is milder than reverse-phase chromatography (RPC), there is still no absolute evidence of measuring the hydrophobicity of native proteins by HIC. In addition, chromatography requires special instruments for the purpose and is usually time consuming and skill requiring. Therefore, it is difficult to use for routine quality-control purposes.

Hydrocarbon binding capacity is rather difficult to relate to similar phenomena in food systems, except for flavor study, e.g., adsorption of aroma compounds with food macromolecules. This minimizes the significance of utilizing this method for hydrophobicity measurement in food research. SDS binding capacity is another empirical method with difficulty in interpretation of the obtained results.

Hydrophobic partition methods are tedious; furthermore, some proteins are insoluble in the nonpolar phase thus floating at the interphase of two separated phases, without attaining equilibrium between the two phases. This restricts the applicability of this type of method in food research.

Triglyceride binding capacity may be the most reasonable method from a practical aspect, because of the importance of lipid-protein interaction on food quality, e.g., emulsion stability, and fat-binding capacity of muscle proteins. However, the GLC or radioactive counting methods of Smith et al.[64] for measuring the bound triglycerides, are complicated, requiring expensive instruments and considerable skill. The fluorescent method of Tsutsui et al.[65] may be preferable as it is much simpler than the procedure of Smith et al.[64] Relatively low-cost fluorometers are commercially available and no professional training is usually required to operate the instruments.

It is probable that hydrophobic fluorescent probe methods will continue to be the most popular methods for a while, especially for quality control, mainly because they are the simplest in analytical procedure and can frequently be completed within 10 min. Furthermore, comparison of the data with other published data is easier for the hydrophobic probe methods due to the prevalence of these methods among protein chemists.

Since there is no absolute method for evaluating the true protein hydrophobicity at the moment, in a practical sense it is valid that the method which can empirically elucidate the function of proteins most satisfactorily could be the preferred method. This situation is quite similar to that of the polarity of compounds and the HLB concept for emulsification. Since there is currently no absolute scale for polarity, empirical scales are being used for elucidating the solvation phenomenon and chromatographic behavior. Similarly, until a better scale appears which is more theoretically valid, the HLB values will be the dominating concept in elucidating the function of surfactants.

Details of the procedures for the empirical methods currently available for protein hydrophobicity measurement, are collated in the Appendix. Intrinsic fluorescence and fluorescence quenching procedural details have not been included due to their restriction to aromatic amino acid behavior rather than overall hydrophobicity.

The published data for protein hydrophobicity are listed in Table 5. However, direct comparison of different methods based on these data may be subject to some inaccuracy and possible criticism, due to the difficulty in obtaining the identical native states for the proteins used for comparison. Probably, comparison of the same proteins under different conditions, e.g., investigation of the effects of processing on the same food proteins, yields more consistent results than comparing the hydrophobicity of different proteins with completely different structures.

Regression analysis was carried out between the hydrophobicity values measured by different methods shown in Table 5, although the sample proteins used for analysis were not always from the same suppliers among researchers. While there was a good correlation between the triglyceride binding capacity (TBC) and CPA hydrophobicity ($r = 0.394$, $P < 0.01$, $n = 56$), the correlation between TBC and ANS was not significant ($r = 0.268$, $P > 0.05$, $n = 35$). This result is contradictory to the result reported by Tsutsui et al.[65] However, this does not mean that there is no significant correlation at all between TBC and ANS hydrophobicities. Tsutsui et al.[65] did not include the unusually large value for BSA in their computation, and it is still possible to obtain a significant correlation by transformation for linearization of ANS hydrophobicity values including that for BSA.

There was a significant correlation between heptane-binding capacity and TBC hydrophobicity ($r = 0.587$, $P < 0.01$, $n = 22$). Also, a significant correlation was found between CPA and HIC hydrophobicities ($r = 0.452$, $P < 0.01$, $n = 55$).

All of these data are supportive for the TBC method, especially the fluorescent procedure, as a reasonable standard method for food protein hydrophobicity until a more reliable method is submitted, hopefully in the near future. CPA hydrophobicity can be an alternative for the TBC method provided it is used above pH 5.0; it is convenient for routine analysis because of its simplicity.

Table 5
PUBLISHED DATA OF PROTEIN HYDROPHOBICITY

Protein	ANS Hayakawa[68]	ANS Tsutsui[65]	CPA Kato[63]	CPA Kato[69]	CPA Hayakawa[68]	CPA Tsutsui[65]	RPC Sadler[51]	RPC Hayakawa[68]	HIC Keshavar[70]	HIC Goheen[63]	HIC Fausnaugh[49]	HIC C$_j$[54]	HIC Phenyl[54]	HIC Gooding[71]	Binding Heptane[60]	Binding SDS[61]	Binding Triglycerides[64]	Binding Triglycerides[65]	Partition Log K	Partition Log K	Partition Zaslavsky[72]
Albumin																					
bovine	1500	—	100	100	100	100	—	8	100	100	100	100	100	100	100	100	100	92	100	100	-27
chicken	10	7	4	1	1	2	—	0	9	86	32	24	43	80	2	14	3	14	25	1	-10
human	—	—					—														-27
Casein																					
α -	85	66	93			25	—	32								30	115	69			
κ -	—	131		13	6	30	—									24	160	140			
Catalase				8																	
α - chymotrypsin	—	0		3		0	—		57	139	81	29	105	120	9		3	38	33		
Chymotrypsinogen A											88	10	13	110	7		2				
Cytochrome C							70			0	3		102	37			8		1	9	-9
Deoxyribonuclease												48	103						11		
Globulin																					
γ - (bovine)				3					79			83	103		7	27	95	47			
7S - (soy)	70	66			6	77	—	13								21		23			36
11S - (soy)	40	52			2	17	—	12									20				
Hemoglobulin	—	—	—			95			108		100	103		10							
Insulin	—	—	—		9							99	102		100	62	130	45	41		
α - lactalbumin	50	91	54			54	90	18	10							22	1		0		
β - lactoglobulin	20	31	9	85	-125	146	100	0	67					113	0	19					
Lipase	—	—		9								40	24				128			78	
Lysozyme	—	—	7				55		83	110	42				24	18	108		56	0	
Myoglobulin	—	—					100		59	39	4				42	18	128				
Ovomucoid																					
chicken	—	—										92	96	71	17		95	29			
turkey	—	—										4	11		29	20	2				
Pepsin	—	1	—	1		2													20	9	-4
Ribonuclease A	—	—					35			59	8	4				13	118	1	27	9	-2

Table 5 (continued)
PUBLISHED DATA OF PROTEIN HYDROPHOBICITY

Protein	ANS	CPA	RPC	HIC
Transferrin, chicken	— —	5 4	— —	16 — 31
Trypsin	— —	6 3	— —	— — —
Trypsin inhibitor	0 1	— 0	1 0	— —

Note: The values of Hayakawa[68] were measured at zero zeta potential. To facilitate comparison, the majority of the data has been standardized relative to a value of 100 for bovine albumin.

a   References are indicated in column heads.

REFERENCES

1. **Rose, G. D., Gierasch, L. M. and Smith, J. A.**, Turns in peptides and proteins, *Adv. Protein Chem.*, 37, 1, 1985.
2. **Nozaki, Y. and Tanford, C.**, The solubility of amino acids and two glycine peptides in aqueous ethanol and dioxane solutions, *J. Biol. Chem.*, 246, 2211, 1971.
3. **Jones, D. D.**, Amino acid properties and side-chain orientation in proteins: a cross correlation approach, *J. Theor. Biol.*, 50, 167, 1975.
4. **Chothia, C.**, The nature of the accessible and buried surfaces in proteins, *J. Mol. Biol.*, 105, 1, 1976.
5. **Levitt, M.**, A simplified representation of protein conformations for rapid simulation of protein folding, *J. Mol. Biol.*, 104, 59, 1976.
6. **Manavalan, P. and Ponnuswamy, P. K.**, Hydrophobic character of amino acid residues in globular proteins, *Nature*, 275, 673, 1978.
7. **Wertz, D. H. and Scheraga, H. A.**, Influence of water on protein structure. An analysis of the preferences of amino acid residues for the inside or outside and for specific conformation in a protein molecule, *Macromolecules*, 11, 9, 1978.
8. **Krigbaum, W. R. and Komoriya, A.**, Local interactions as a structure determinant for protein molecules: II, *Biochim. Biophys. Acta*, 576, 204, 1979.
9. **Janin, J.**, Surface and inside volumes in globular proteins, *Nature*, 277, 491, 1979.
10. **Robson, B. and Osguthorpe, D. J.**, Refined models for computer simulation of protein folding. Applications to the study of conserved secondary structure and flexible hinge points during the folding of pancreatic trypsin inhibitor, *J. Mol. Biol.*, 132, 19, 1979.
11. **Meirovitch, H., Rackovsky, S., and Scheraga, H. A.**, Empirical studies of hydrophobicity. I. Effect of protein size on the hydrophobic behavior of amino acids, *Macromolecules*, 13, 1398, 1980.
12. **Nishikawa, K. and Ooi, T.**, Prediction of the surface-interior diagram of globular proteins by an empirical method, *Int. J. Peptide Protein Res.*, 16, 19, 1980.
13. **Wolfenden, R., Andersson, L., Cullis, P. M., and Southgate, C. C. B.**, Affinities of amino acid side chains for solvent water, *Biochemistry*, 20, 849, 1981.
14. **Yunger, L. M. and Cramer, R. D.**, III. Measurement and correlation of partition coefficients of polar amino acids, *Mol. Pharm.*, 20, 602, 1981.
15. **Kyte, J. and Doolittle, R. F.**, A simple method for displaying the hydropathic character of a protein, *J. Mol. Biol.*, 157, 105, 1982.
16. **Sweet, R. M. and Eisenberg, D.**, Correlation of sequence hydrophobicities measures similarity in three-dimensional protein structure, *J. Mol. Biol.*, 171, 479, 1983.
17. **Rose, G. D., Geselowitz, A. R., Lesser, G. J., Lee, R. H., and Zehfus, M. H.**, Hydrophobicity of amino acid residues in globular proteins, *Science*, 229, 834, 1985.
18. **Lee, B. and Richards, F. M.**, The interpretation of protein structures: estimation of static accessibility, *J. Mol. Biol.*, 55, 379, 1971.
19. **Burstein, E. A., Vedenkina, N. S., and Ivkova, M. N.**, Fluorescence and the location of tryptophan residues in protein molecules, *Photochem. Photobiol.*, 18, 263, 1973.
20. **Morrisett, J. D., Sparrow, J. T., Hoff, H. F., and Gotto, A. M. Jr.**, Methods for studying lipid-protein interactions, *Cardiovasc. Res. Center Bull.*, 12, 39, 1973.
21. **Dufourcq, J., Faucon, J. F., Lussan, C., and Bernon, R.**, Study of lipid-protein interactions in membrane models: intrinsic fluorescence of cytochrome b-phospholipid complexes, *FEBS Lett.*, 57, 112, 1975.
22. **Genot, C., Montenay-Garestier, T., and Drapron, R.**, Intrinsic spectrofluorometry applied to soft wheat (*Tricum aestivum*) flour and gluten to study lipid-protein interactions, *Lebensm.-Wiss. u. -Technol.*, 17, 129, 1984.
23. **Altekar, W.**, Fluorescence of proteins in aqueous neutral salt solutions. I. Influence of anions, *Biopolymers*, 16, 341, 1977.
24. **Eftink, M. R. and Ghiron, C. A.**, Exposure of tryptophanyl residues and protein dynamics, *Biochemistry*, 16, 5546, 1977.
25. **Eftink, M. R., Zajicek, J. L., and Ghiron, C. A.**, A hydrophobic quencher of protein fluorescence: 2,2,2-trichloroethanol, *Biochim. Biophys. Acta*, 491, 473, 1977.
26. **Turner, D. C. and Brand, L.**, Quantitative estimation of protein binding site polarity. Fluorescence of *N*-arylaminonaphthalenesulfonates, *Biochemistry*, 7, 3381, 1968.
27. **Terada, H., Hiramatsu, K., and Aoki, K.**, Heat denaturation of serum albumin monitored by 1-anilinonaphthalene-8-sulfonate, *Biochim. Biophys. Acta*, 622, 161, 1980.
28. **Daban, J. R. and Guasch, M. D.**, Exposed hydrophobic regions in histone oligomers studied by fluorescence, *Biochim. Biophys. Acta*, 625, 237, 1980.
29. **Palumbo, G. and Ambrosio, G.**, The use of 1-anilinonaphthalene-8-sulfonate (ANS) for studying the effects of iodination on thyroglobulin conformation, *Arch. Biochem. Biophys.*, 212, 37, 1981.

30. **Penzer, G.,** 1-Anilinonaphthalene-8-sulphonate. The dependence of emission spectra on molecular conformation studied by fluorescence and proton-magnetic resonance, *Eur. J. Biochem.,* 25, 218, 1972.

31. **Sklar, L. A., Hudson, B. S., and Simoni, R. D.,** Conjugated polyene fatty acids as fluorescent probes: binding to bovine serum albumin, *Biochemistry,* 16, 5100, 1977.

32. **Tecoma, E. S., Sklar, L. A., Simoni, R. D., and Hudson, B. S.,** Conjugated polyene fatty acids as fluorescent probes: biosynthetic incorporation of parinaric acid by Escherichia coli and studies of phase transitions, *Biochemistry,* 16, 829, 1977.

33. **Fraley, R., Jameson, D. M., Kaplan, S.,** The use of the fluorescent probe α-parinaric acid to determine the physical state of the intracytoplasmic membranes of the photosynthetic bacterium, *Rhodopseudomonas sphaeroides, Biochim. Biophys. Acta,* 511, 52, 1978.

34. **Barrow, D. A. and Lentz, B. R.,** Membrane structural domains. Resolution limits using diphenylhexatriene fluorescence decay, *Biophys. J.,* 48, 221, 1985.

35. **Babel, A. O., Turner, M., and Dolmans, M.,** Spectral studies of the interaction of bovine α-lactalbumin and egg-white lysozyme with 2-*p*-toluidinylnaphthalene-6-sulfonate, *Eur. J. Biochem.,* 30, 26, 1972.

36. **Bohnert, J. L., Malencik, D. A., Anderson, S. R., Teller, D., and Fischer, E. H.,** Binding of 5,5′-*bis* 8-(phenylamino)-1-naphthalenesulfonate by the regulatory subunits of adenine cyclic 3′.5′-phosphate dependent protein kinase, *Biochemistry,* 21, 5570, 1982.

37. **Rekker, R. F.,** in *The Hydrophobic Fragmental Constant,* Rekker, R. F., Ed., Vol. 1, Elsevier, New York, 1977.

38. **Jinno, K.,** Utilization of micro-HPLC for the direct measurement of the partition coefficients needed in studying the quantitative structure-activity relationships, *Chromatographia,* 15, 723, 1982.

39. **Hjerten, S.,** The synthesis and the use of some alkyl aryl derivatives of agarose, *J. Chromatog.,* 101, 281, 1974.

40. **Kissing, W. and Reiner, R. H.,** Chromatographic investigation into the interaction between proteins and polymer matrices. I. Agarose gels with substituents of different length and hydrophobicity, *Chromatographia,* 10, 129, 1977.

41. **Yamashiro, D.,** The purification of peptides by partition chromatography based on a hydrophobicity scale, *Int. J. Peptide Protein Res.,* 13, 5, 1979.

42. **O'Hare, M. J. and Nice, E. C.,** Hydrophobic high-performance liquid chromatography of hormonal polypeptides and proteins on alkylsilane-bonded silica, *J. Chromatog.,* 171, 209, 1979.

43. **Meek, J. L.,** Prediction of peptide retention times in high-pressure liquid chromatography on the basis of amino acid composition, *Proc. Natl. Acad. Sci. U.S.A.,* 77, 1632, 1980.

44. **Imoto, T. and Okazaki, K.,** A simple peptide fractionation by hydrophobic chromatography with a prepacked reversed-phase column, *J. Biochem.,* 89, 437, 1981.

45. **Wilson, K. J., Honegger, A., Stotzel, R. P., and Hughes, G. J.,** The behavior of peptides on reverse-phase supports during high-pressure liquid chromatography, *Biochem. J.,* 199, 31, 1981.

46. **Bigelow, C. C. and Chapman, M.,** in *Handbook of Biochemistry and Molecular Biology,* 1, 209, 1976. Cited by Wilson et al. 45.

47. **Pliska, V. and Fauchere, J.-L.,** in *Peptides,* Gross, E. and Meienhofer, J., Eds., Pierce Chemical, Rockford, Ill., 1979.

48. **Segrest, J. P. and Feldmann, R. J.,** Membrane proteins: amino acid sequence and membrane penetration, *J. Mol. Biol.,* 87, 853, 1974.

49. **Faunaugh, J. L., Kennedy, L. A., and Regnier, F. E.,** Comparison of hydrophobic-interaction and reversed-phase chromatography of proteins, *J. Chromatog.,* 317, 141, 1984.

50. **Luiken, J., van der Zee, R., and Welling, G. W.,** Structure and activity of proteins after reversed-phase high-performance liquid chromatography, *J. Chromatog.,* 284, 482, 1984.

51. **Sadler, A. J., Micanovic, R., Katzenstein, G. E., Lewis, R. V., and Middaugh, C. R.,** Protein conformation and reversed-phase high-performance liquid chromatography, *J. Chromatog.,* 317, 93, 1984.

52. **Ingraham, R. H., Lau, S. Y. M., Taneja, A. K., and Hodges, R. S.,** Denaturation and the effects of temperature on hydrophobic-interaction and reversed-phase high-performance liquid chromatography of proteins, Bio-gel TSK-phenyl-5-PW column, *J. Chromatog.,* 327, 77, 1985.

53. **Goheen, S. C. and Engelhorn, S. C.,** Hydrophobic interaction high-performance liquid chromatography of proteins, *J. Chromatog.,* 317, 55, 1984.

54. **Hofstee, B. H. and Otillio, N. F.,** Modifying factors in hydrophobic protein binding by substituted agaroses, *J. Chromatog.,* 161, 153, 1978.

55. **Maier, H. G.,** Zur Bindung fluchtiger Aromastoffe und Lebensmittel. II. Extraktor-Methode, *Z. Lebensm.-Unters. u. -Forsch.,* 141, 332, 1970.

56. **Maier, H. G.,** Zur Bindung fluchtiger Aromastoffe an Proteine, *Dtsch. Lebensm.-Rundschau,* 70, 349, 1974.

57. **Maier, H. G.,** Binding of volatile aroma substances to nutrients and foodstuffs, *Proc. Int. Symp. Aroma Res. Zeist.,* 143, 1975.

58. **Bigelow, C. C.**, On the average hydrophobicity of proteins and the relation between it and protein structure, *J. Theoret. Biol.*, 16, 187, 1967.
59. **Mohammadzadeh-K, A., Feeney, R. E., and Smith, L. M.**, Hydrophobic binding of hydrocarbons by proteins. I. Relationship of hydrocarbon structure, *Biochem. Biophys. Acta*, 194, 246, 1969.
60. **Mohammadzadeh-K, A., Smith, L. M., and Feeney, R. E.**, II. Relationship of protein structure, *Biochim. Biophys. Acta*, 194, 256, 1969.
61. **Reynolds, J. A. and Tanford, C.**, Binding of dodecyl sulfate to proteins at high binding ratios. Possible implications for the state of proteins in biological membranes, *Proc. Natl. Acad. Sci. U.S.A.*, 66, 1002, 1970.
62. **Takenaka, O., Aizawa, S., Tamura, Y., Hirano, J., and Inada, Y.**, Sodium dodecylsulfate and hydrophobic regions in bovine serum albumin, *Biochim. Biophys. Acta*, 263, 696, 1972.
63. **Kato, A., Matsuda, T., Matsudomi, N., and Kobayashi, K.**, Determination of protein hydrophobicity using a sodium dodecyl sulfate binding method, *J. Agric. Food Chem.*, 32, 284, 1984.
64. **Smith, L. M., Fantozzi, P., and Creveling, R. K.**, Study of triglyceride-protein interaction using a microemulsion-filtration method, *J. Am. Oil Chem. Soc.*, 60, 960, 1983.
65. **Tsutsui, T., Li-Chan, E., and Nakai, S.**, A simple fluorometric method for fat-binding capacity as an index of hydrophobicity of proteins, *J. Food Sci.*, 51, 1268, 1986.
66. **Shanbhag, V. P. and Axelsson, C.-G.**, Hydrophobic interaction determined by partition in aqueous two-phase systems. Partition of proteins in systems containing fatty-acid esters of poly(ethylene glycol), *Eur. J. Biochem.*, 60, 17, 1975.
67. **Zaslavsky, B. Y., Miheeva, L. M., and Rogozhin, S. V.**, Relative hydrophobicity of surfaces of erythrocytes from different species as measured by partition in aqueous two-polymer phase systems, *Biochim. Biophys. Acta*, 588, 89, 1979.
68. **Hayakawa, S. and Nakai, S.**, Relationships of hydrophobicity and net charge to the solubility of milk and soy proteins, *J. Food Sci.*, 50, 486, 1985.
69. **Kato, A. and Nakai, S.**, Hydrophobicity determined by a fluorescence probe method and its correlation with surface properties of proteins, *Biochim. Biophys. Acta*, 624, 13, 1980.
70. **Keshavarz, E. and Nakai, S.**, The relationship between hydrophobicity and interfacial tension of proteins, *Biochim. Biophys. Acta*, 576, 269, 1979.
71. **Gooding, D., Schmuck, M. N., and Gooding, K. M.**, Analysis of proteins with new, mildly hydrophobic high-performance liquid chromatography packing materials, *J. Chromatog.*, 296, 107, 1984.
72. **Zaslavsky, B. Y., Mestechkina, N. M., and Rogozhin, S. V.**, Characteristics of protein-aqueous medium interactions measured by partition in aqueous ficol-dextran biphasic system, *J. Chromatog.*, 260, 329, 1983.

Chapter 3

HYDROPHOBICITY-FUNCTIONALITY RELATIONSHIP OF FOOD PROTEINS

TABLE OF CONTENTS

I. PROTEIN SOLUBILITY

Solubility of proteins has been thought to be the most important factor for their functionality, and thus, their potential applications. For example, different uses of soy flour based on their nitrogen solubility index (NSI) values have been suggested.[1] Soy flour which has received the minimum heat treatment may have an NSI of more than 90%, but the NSI of soy flour can range from 10 to 90%. However, due to the empirical nature of these values, food processors are recommended to define the NSI range of the soy flour or grits which gives the optimum results for their food product.

It is agreed that surfactant properties are related to the aqueous solubility of proteins.[2] In the extraction and precipitation of proteins during processing, information on changes in solubility is of prime importance.[3] Good solubility can markedly expand potential applications of proteins. Denaturation of proteins, especially when caused by heating, frequently impairs their functional properties and is expressed as a loss of solubility.[1]

However, many authors have reported that emulsifying properties and solubility are not closely correlated.[4,5] Including hydrophobicity parameters in addition to solubility as predictor variables in multiple regression analysis for protein functionality, improved the coefficients of multiple determination.[6-9] For example, when the emulsifying ability of meat proteins (n = 26) was investigated, the R^2 value was improved from 0.730 to 0.917 before and after including *cis*-parinarate (CPA) hydrophobicity data in addition to solubility data in regression analysis.[10] Li-Chan et al.[10] also reported that hydrophobicity parameters were influential in emulsifying properties of meat proteins with low solubility.

According to Bigelow,[11] proteins with high hydrophobicity and low charge frequency are more soluble in the organic phase than aqueous phase, while proteins with low hydrophobicity and high charge frequency are more soluble in the aqueous phase. Charge frequency and hydrophobicity are likely to be two structural features having the greatest influence on the solubility of protein.

A. Solubility and Hydrophobicity of Proteins

In order to study the solubility-hydrophobicity relationship of proteins without interference from charge frequency effects, hydrophobicity and solubility were measured for proteins under conditions of zero zeta potential.[12] Forty-two milk and soy protein samples were used for the study, including 11 native samples and 31 partially denatured samples as described in the legend to Figure 1. Insolubility was calculated as 100% × (total protein-supernatant protein)/total protein. The supernatant was obtained after centrifuging 0.1 to 0.2% protein solutions at 6000 × g for 30 min.

CPA as well as diphenylhexatriene (DPH) hydrophobicity did not show significant relationship with insolubility measured at the zeta potential of zero mV (Figure 1). Anilino-naphthalene sulfonate (ANS) hydrophobicity showed a good relationship with insolubility (Figure 2) with r = 0.592 (P <0.001), while hydrophobicity expressed as percent isopropanol required to elute proteins from a phenyl Sepharose® CL-4B (PSC) column showed a more highly significant relationship (Figure 3) with r = 0.775 (P <0.001).

B. Correlation of Hydrophobicity and Zeta Potential to Protein Solubility

Another important factor affecting solubility is charge frequency of protein molecules. Proteins show minimum solubility at zero zeta potential. Backward stepwise multiple regression analysis was applied to 189 data of solubility, measured at different zeta potentials ranging from 0 to 30 mV, as a function of hydrophobicity and the absolute value of zeta potential (ZP). With CPA hydrophobicity, contribution of its deviation to the solubility difference was slight. With ANS hydrophobicity, a significant relationship (P <0.001) was obtained with the regression equation of:

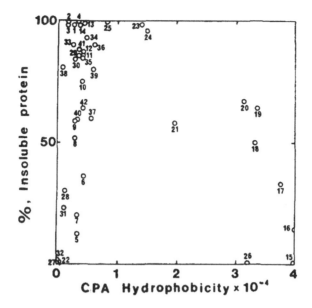

FIGURE 1. Relationship between protein insolubility at zero zeta potential and hydrophobicity measured fluorometrically using *cis*-parinaric acid (CPA). (1) α-casein; (2) α_{s1}-casein; (3) β-casein; (4) κ-casein; (5) native α-lactalbumin (α-La); (6) α-La heated at 90°C; (7) — (13) α-La heated at 30, 40, 45, 50, 60, 80, and 90°C in the presence of 10 mM mercaptoethanol (ME); (14) α-La treated with 8 M urea and 100 mM ME; (15) native β-lactoglobulin (β-Lg); (16) — (20) β-Lg heated at 75, 80, 85, 90, and 95°C; (21) β-Lg treated with urea and ME; (22) native ovalbumin; (23) ovalbumin heated at 85°C; (24) bovine serum albumin (BSA) treated with urea and ME; (25) BSA heated at 90°C; (26) native BSA; (27) native soybean trypsin inhibitor (STI); (28) STI heated at 95°C; (29) STI heated at 95°C in the presence of ME; (30) STI treated with urea and ME; (31) STI heated at 95°C at pH 2; (32) STI treated with alkali (pH 12); (33) native soy 7S globulin (7S); (34) 7S heated at 95°C; (35) 7S treated with urea and ME; (36) 7S heated at 95°C at pH 2; (37) 7S treated with alkali; (38) native soy 11S globulin (11S); (39) 11S heated at 95°C; (40) 11S treated with urea and ME; (41) 11S heated at 95°C at pH 2; (42) 11S treated with alkali. (From Hayakawa, S. and Nakai, S., *J. Food Sci.*, 50, 486, 1985. With permission.)

$$\% \text{ insoluble protein} = 0.78 \text{ ANS} - 0.0014 \text{ ANS}^2 - 0.010 \text{ ANS} \cdot \text{ZP}$$

$$- 0.042 \text{ ZP}^2 + 13.4 \ (R^2 = 0.480) \tag{1}$$

With PSC hydrophobicity, an even higher coefficient of determination (R^2) was obtained ($P < 0.001$), with the regression equation of:

$$\% \text{ insoluble protein} = 3.43 \text{ PSC} - 0.057 \text{ PSC} \cdot \text{ZP}$$

$$- 0.042 \text{ ZP}^2 + 4.44 \ (R^2 = 0.612) \tag{2}$$

According to Burley and Sleigh,[13] the elution position during hydrophobic chromatography depends on the hydrophobicity alone when the proteins are similar in structure, and proteins with many aromatic amino acids on the surface may bind well to phenyl Sepharose® through $\pi - \pi$ bond interaction. This may indicate the importance of aromatic hydrophobicity in protein solubility as discussed later. Figure 4 shows a three-dimensional plot illustrating the relationship of PSC hydrophobicity and absolute value of zeta potential with protein insolubility. It is clear that the protein insolubility increases with an increase of hydrophobicity and with a decrease of zeta potential of the protein.

Since there has been no strong correlation of CPA hydrophobicity with solubility, a combination of CPA hydrophobicity and solubility can be justified when used in multiple regression analysis for describing protein functionality, with little risk of multicollinearity of the descriptor variables.

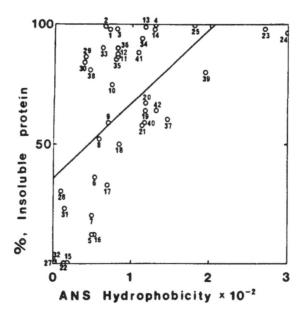

FIGURE 2.   Relationship between protein insolubility at zero zeta potential and hydrophobicity measured fluorometrically using 1-anillino-8-naphthalenesulfonate (ANS). Sample identification is the same as in Figure 1. Regression equation: [% insoluble protein] = 36.2 + 0.309 ANS (r = 0.592, P <0.001). (From Hayakawa, S. and Nakai, S., *J. Food Sci.*, 50, 486, 1985. With permission.)

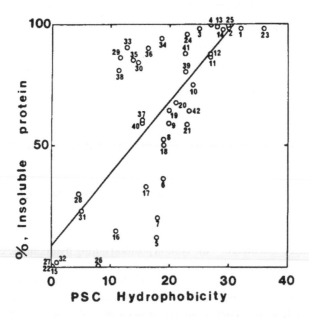

FIGURE 3.   Relationship between protein insolubility at zero zeta potential and hydrophobicity measured by the hydrophobic chromatography on a Phenyl Sepharose® CL-4B column. Sample identification is the same as in Figure 1. Regression equation: [% insoluble protein] = 9.30 + 2.93 PSC (r = 0.775, P <0.001). (From Hayakawa, S. and Nakai, S., *J. Food Sci.*, 50, 486, 1985. With permission.)

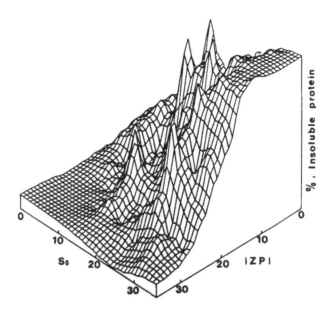

FIGURE 4. Three-dimensional surface plot of protein insolubility as functions of zeta potential (ZP) and PSC hydrophobicity (S_o). Protein insolubility was measured at different zeta potentials ranging from 0 to 30 mV. (From Hayakawa, S. and Nakai, S., *J. Food Sci.*, 50, 486, 1985. With permission.)

C. Importance of Aromatic Hydrophobicity

CPA possesses an aliphatic hydrocarbon chain, while ANS is composed of aromatic rings. Aromatic groups are attached to agarose in phenyl Sepharose®. It is likely that the binding sites for CPA on protein molecules differ from the sites for ANS and PSC. The fact that hydrophobicities measured with ANS and PSC showed significant relationship with protein solubility while hydrophobicity measured with CPA did not, suggested possible differences in properties of hydrophobicity originating from aliphatic amino acid side chains (e.g., Val, Leu and Ile) and hydrophobicity originating from aromatic amino acid side chains (e.g., Phe, Trp and Tyr) tentatively designated "aliphatic hydrophobicity" and "aromatic hydrophobicity", respectively.

Differences in behavior between aliphatic and aromatic amino acids have been exemplified in the specific adsorption of aromatic amino acids on crosslinked dextran gel from a mixture of amino acids including aliphatic amino acids,[14] and in the preferential binding of some proteins by aliphatic adsorbents and other proteins by aromatic adsorbents.[15] According to Snyder,[16] not only the polarity scale (which reflects overall solvation capacity) but also the selectivity (which is relevant to the chemical structure) are essential for classifying the solvents. The π^* scale of Kamlet et al.,[17] another polarity scale, is usually greater for aromatic solvents than expected from aliphatic solvents. This is ascribed to the higher polarizability of the aromatic solvents.

Aromatic amino acid residues in thermophilic enzymes were reported as having a more nonpolar microenvironment than those in the respective mesophilic enzymes.[18,19] It is suggested that, although they are more hydrophobic in nature, aromatic residues are not buried inside the molecules due to their bulky structure. On the other hand, less hydrophobic aliphatic amino acid residues find their way inside the protein on folding as they are smaller in size and possess greater elasticity.[20] Burley and Petsko[21] proposed the possible importance of aromatic-aromatic interaction in protein stabilization based on examination of 590 pairs of neighboring aromatic side chains in 34 known protein structures. Since the structure definitely relates to the functionality of proteins, it is reasonable to distinguish aromatic hydrophobicity from aliphatic hydrophobicity, because of the characteristic role which may be played by aromatic amino acid residues in protein structure.

II. EMULSIFICATION

For the preparation of food emulsions, surface active agents are widely used. For the selection of proper surfactants, the HLB (hydrophile lipophile balance) scale is commonly used. As described before, the HLB concept is another expression of polarity and defined as "one-fifth of the percent hydrophilic fraction in the surfactant molecule" as expressed as:

$$HLB = (\text{weight of hydrophilic})/(\text{weight of total}) \times 100/5$$

The factor 5 is purely empirical, thus HLB values for 100% lipophilic and hydrophilic molecules are 0 and 20, respectively. An example of computation of the HLB value is shown for food emulsifiers with known chemical structure (Figure 5). HLB value and protein hydrophobicity are similar in concept with the relationship: [protein hydrophobicity] = 1 − [HLB], since the protein hydrophobicity is defined as [nonpolar residues] / ([nonpolar] + [polar]) and HLB as [polar radicals]/ ([polar] + [nonpolar]).

Selection of proper emulsifiers for preparation of a food emulsion is made by choosing an emulsifier with the HLB value which matches the required HLB value of the oil or lipid to be emulsified based on an assumption that the HLB is additive. The HLB approach functions well for oils composed of single triglycerides, e.g., olive oil; however, the HLB value of emulsifier is rather broad and undefined for making the most stable emulsion in the case of oils composed of mixed triglycerides, e.g., butter.

As surfactants are compounds which contain both hydrophilic and lipophilic radicals in the same molecules, protein is also a surface active compound since it is composed of hydrophilic as well as hydrophobic amino acid residues. Therefore, protein hydrophobicity should explain its emulsifying capacity. The ability of protein to aid in the formation and stabilization of emulsions is important for preparation of chopped or comminuted meats, cake batters, coffee whiteners, milk, mayonnaise, salad dressings, and frozen desserts.

A significant correlation ($P < 0.01$) was obtained between the emulsifying capacity (as well as interfacial tension) and hydrophobicity of proteins determined fluorometrically by Kato and Nakai as shown in Figure 6.[22] A relative hydrophobicity measured by the method of Kato and Nakai[22] using *cis*-parinarate (CPA) could be a measure of "surface hydrophobicity", which is not necessarily directly correlated with the total content of hydrophobic amino acids in the protein molecule. Hydrophobic sites on the surface of the protein molecules could be a critical factor for good emulsification. Nakai et al.[23] reported that the surface hydrophobicity showed good correlations with surface tension, interfacial tension, and emulsifying activity of soy, canola, and sunflower proteins and their derivatives treated with a linoleate and proteinase.

For emulsification, surface active agents should migrate to the oil/water interface, thus decreasing the interfacial tension from 14 to 20 dyn/cm without surfactants to below 10 dyn/cm. To migrate towards the interface, a protein molecule should be mobile. Since good mobility cannot be expected from coagulated or insoluble proteins, good solubility of the proteins is also essential for formation of good emulsions.

Emulsifying activity index (EAI, m²/g) and emulsion stability index (ESI, min) of Pearce and Kinsella[24] were measured for native and denatured samples of soy, pea, canola, sunflower, gluten, casein, whey proteins, β-lactoglobulin, gelatin, myosin, ovalbumin, and BSA.[6] The regression equations obtained were:

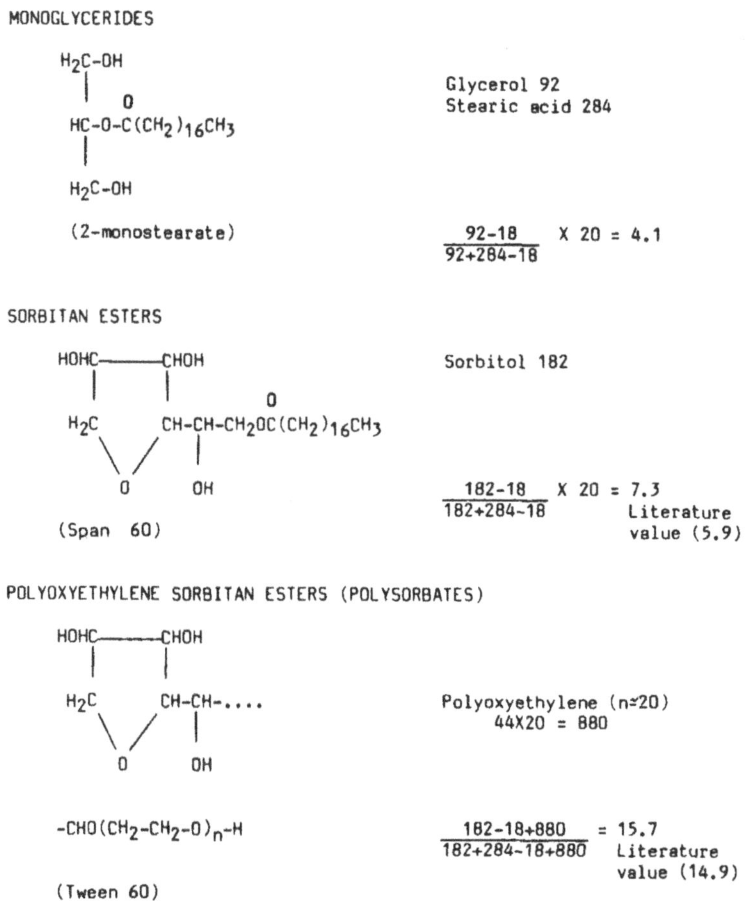

MONOGLYCERIDES

```
H2C-OH
 |
 |      O
HC-O-C(CH2)16CH3
 |

H2C-OH
```

Glycerol 92
Stearic acid 284

(2-monostearate)

$$\frac{92-18}{92+284-18} \times 20 = 4.1$$

SORBITAN ESTERS

```
HOHC————CHOH
  |      |
  |      |        O
 H2C    CH-CH-CH2OC(CH2)16CH3
  \    / |
   \  /  OH
    O
```

Sorbitol 182

(Span 60)

$$\frac{182-18}{182+284-18} \times 20 = 7.3$$
Literature
value (5.9)

POLYOXYETHYLENE SORBITAN ESTERS (POLYSORBATES)

```
HOHC————CHOH
  |      |
  |      |
 H2C    CH-CH-....
  \    / |
   \  /  OH
    O
```

Polyoxyethylene (n≈20)
44X20 = 880

```
-CHO(CH2-CH2-O)n-H
```

$$\frac{182-18+880}{182+284-18+880} = 15.7$$
Literature
value (14.9)

(Tween 60)

FIGURE 5. Computation of HLB values based on chemical structure of synthetic emulsifiers. Because the commercial products are not 100% pure, the theoretical calculation
is not the same as the published data for the commercial products.

$$EAI = 16.877 + 0.214S_o + 0.932s - 0.007s^2 \quad (n = 52, R^2 = 0.583, P < 0.001)$$

$$ESI = -69.463 + 0.565S_o + 2.034s - 0.004S_o \cdot s - 0.012s^2$$

$$(n = 49, R^2 = 0.584, P < 0.001) \tag{3}$$

where S_o (%$^{-1}$) and s (%) are CPA hydrophobicity and solubility index, respectively.

Response surface plots (contours) to visualize the effects of S_o and solubility on the
emulsifying ability are shown in Figure 7. Regardless of the solubility index, as hydrophobicity increases, both EAI and ESI initially increase and then decline. This is reasonable
considering the matching requirements between HLB and required HLB for emulsification
or between solute and solvent in terms of polarity. The emulsifying properties of proteins
ultimately depend on the balance between the hydrophile and lipophile, and do not necessarily
increase linearly as the proteins become more hydrophobic.

A similar observation was made by Aoki et al.;[25] excessive denaturation of soy protein
by n-propanol resulted in a lower emulsion stability as compared to the moderate denaturation
by ethanol. If this is the case, the relationship between emulsifying properties and hydrophobicity may or may not be linear depending on the range being covered on the hydrophobicity scale.

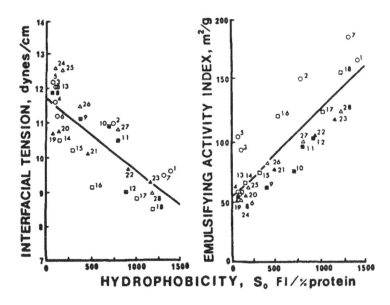

FIGURE 6.   Relationships of S_o with interfacial tension and emulsifying activity of proteins. 1, bovine serum albumin; 2, β-lactoglobulin; 3, trypsin; 4, ovalbumin; 5, conalbumin; 6, lysozyme; 7, κ-casein; 8, 9, 10, 11, and 12, denatured ovalbumin by heating at 85°C for 1, 2, 3, 4, and 5 min, respectively; 13, 14, 15, 16, 17, and 18, denatured lysozyme by heating at 85°C for 1, 2, 3, 4, 5, and 6 min, respectively; 19, 20, 21, 22, and 23, ovalbumin bound with 0.2, 0.3, 1.7, 5.7, and 7.9 mol of dodecyl sulfate/mol of protein, respectively; 24, 25, 26, 27, and 28, ovalbumin bound with 0.3, 0.9, 3.1, 4.8, and 8.2 mol of linoleate/mol of protein, respectively. Interfacial tension: measured at corn oil/0.2% protein interface with a Fisher Surface Tensiomat Model 21. Emulsifying activity index: calculated from the absorbance at 500 nm of the supernatant after centrifuging blended mixtures of 2 mℓ of corn oil and 6 mℓ of 0.5% protein in 0.01 M phosphate buffer, pH 7.4. S_o: initial slope of fluorescence intensity (FI) vs. percent protein plot. 10 μℓ of 3.6 mM *cis*-parinaric acid solution was added to 2 mℓ of 0.002 to 0.1% protein in 0.01 M phosphate buffer, pH 7.4, containing 0.002% NaDodSO₄. FI was measured at 420 nm by exciting at 325 nm. (Adapted from Kato and Nakai.[22])

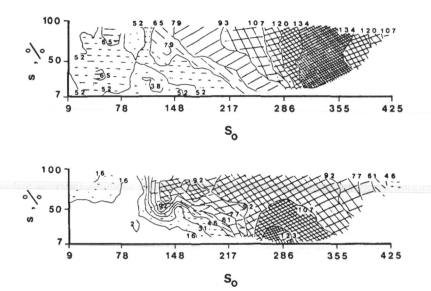

FIGURE 7.   Upper: emulsifying-activity index response surface contour as a function of hydrophobicity (S_o) and solubility index(s). Lower: emulsion-stability index response surface contour as a function of hydrophobicity (S_o) and solubility index(s). (From Voutsinas, L. P. et al., *J. Food Sci.*, 48, 26, 1983. With permission.)

III. FOAMING

Protein foams are important for preparation of several food products, e.g., meringues, soufflés, whipped toppings, chiffon desserts, icings, fudges, and bakery products. Due to the importance of protein foams in the food industry, a large body of studies has been carried out to investigate the foaming behavior of proteins under a variety of conditions.[26-28] These studies have demonstrated that numerous factors including pH, temperature, the presence of salts, sugars, and lipids, and the protein source, affect the foaming behavior of proteins. Many other studies have concentrated on the relationships between foaming properties and the physical and chemical properties of surfaces, e.g., surface pressure, surface protein concentration, surface elasticity, and surface viscosity.[29-31] According to Phillips[30] the dynamic dilational modulus:

$$\epsilon = -A(d\pi/dA) \tag{4}$$

which is the change in the surface pressure ($d\pi$) caused by a relative change in surface area (dA/A), is important for foamability as well as foam stability. Since the surface pressure is the decrease in surface tension produced by adsorbed protein films at the surface and since the adsorption is due to their amphiphilic nature, a close relation between the hydrophobicity and foaming ability of proteins can be expected. Horiuchi et al.[32] correlated the foam stability of five enzyme-hydrolyzed proteins with the content of surface hydrophobic regions of the molecules, which was fluorometrically measured using ANS as a hydrophobic probe (Figure 8).

A number of workers have shown that protein solubility makes an important contribution to the foaming behavior of proteins. Hermansson et al.[33] found that the foaming capacity of fish protein concentrate was provided by the soluble protein, which accounted for 1% of the total protein. Wang and Kinsella[34] found that for alfalfa leaf protein the pH-foaming capacity curve paralleled its pH-solubility profile.

When a method similar to that of Waniska and Kinsella[35] was used for determining foaming capacity (FC) and foam stability (FS) for 11 to 19 different proteins which were relatively pure and undenatured, the regression equations obtained were:[8]

$$FC = 260.4 \ln S_e - 0.301 S_e + 158.6 \ln d - 2.724d - 1820 \ (R^2 = 0.772;$$

$$n = 19; \quad P < 0.01)$$

$$FC = 1493\eta + 25.93 \ln S_e - 1775 \ (R^2 = 0.779, P < 0.01)$$

$$\ln (FS + 0.18) = 0.0102 \ H\phi_{avg} - 9.791 \ (r = 0.807, n = 11, P < 0.01) \tag{5}$$

where S_e, d and η(mPa·s) are S_o of unfolded protein samples, dispersibility, and viscosity, respectively; $H\phi_{avg}$ is the average hydrophobicity calculated according to Bigelow,[11] implicating that more exposure of hydrophobic sites is desirable for good foam stability. These relationships are illustrated in Figure 9.

Exposable hydrophobicity (S_e) rather than surface hydrophobicity (S_o) showed better correlations to foaming capacity, unlike emulsifying properties, probably indicating more extensive uncoiling of the protein molecules at an air/water interface than at an oil/water interface as reported by Graham and Phillips.[36]

Because large amounts of protein samples were required for foaming test compared to emulsifying test, e.g., EAI of Pearce and Kinsella,[24] the number of proteins used in this study were restricted. This is the reason why more independent variables could not be used simultaneously in the multiple regression analysis. By using denatured samples, or chemically

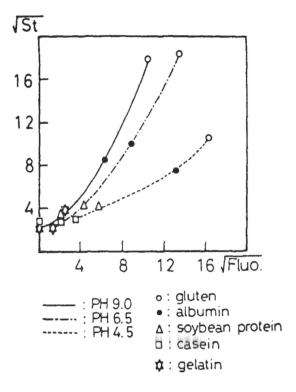

FIGURE 8.   Correlation between foam stability (S_t) and content of the hydrophobic region on the molecular surface (relative fluorescence $=$ Fluo.) From Horiuchi, T. et al., *Food Chem.*, 3, 35, 1978. With permission.)

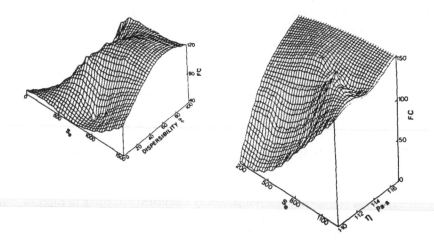

FIGURE 9.   Left: Relationship of hydrophobicity and solubility with foaming capacity of proteins. Proteins: bovine serum albumin, ovalbumin, lysozyme, β-lactoglobulin, ovomucoid, trypsin, ribonuclease A, conalbumin, β-casein, and κ-casein. S_e: S_o measured after 1.5% protein solutions with 1.5% NaDodSO₄ were heated for 10 min in boiling water and diluted. Solubility: percent Kjeldahl N of the supernatant in total N after centrifuging 1% protein in 0.1 M phosphate buffer, pH 7.0, stirred on a magnetic stirrer for 10 min. Foaming capacity: volume (mℓ) of foam from 15 mℓ of 0.1% protein solution by air sparging through a sintered glass disk. The three-dimensional plot was drawn by an Amdahl 470 V/8 computer by using the UBC SURFACE program. Right: Relationship of hydrophobicity and viscosity to foaming capacity of proteins. Viscosity: measured by using an Ostwald viscometer. (From Townsend, A.-A. and Nakai, S., *J. Food Sci.*, 48, 588, 1983. With permission.)

or enzymatically modified samples to increase the number of samples, it may be possible to include more independent variables in the regression equations, thus the contributions of a variety of factors, e.g., hydrophobicity, solubility, viscosity, etc., could be more precisely defined. A simpler form of correlation with hydrophobicity alone, as found by Horiuchi et al.,[32] may be interpreted as an example of eliminating the solubility effect by using partially hydrolyzed proteins, as all of the samples are fully soluble.

IV. THERMAL FUNCTIONAL PROPERTIES

Bigelow[11] suggested that thermal stability might be attributed to the hydrophobic effects in thermophilic proteins compared with their mesophilic counterparts. The importance of hydrophobic interaction in determining protein stability was emphasized since their strength was supposed to increase with temperature up to 60 to 70°C, whereas the contribution of hydrogen bonding to the free energy was said to remain more or less constant with increasing temperature.[37] This favors the idea that hydrophobic force is the dominant factor in the thermal stability of proteins. However, an extensive comparison of the proportions of hydrophobic residues in thermophilic and in mesophilic proteins, has not revealed any such definite relationship between hydrophobicity and thermostability.[38,39] Ikai[40] has proposed that an aliphatic index has a better quantitative correlation with the thermostability of globular proteins.

Ponnuswamy et al.[41] investigated the relationship between the thermal stability of 15 globular proteins and their amino acid compositions and found that groups of residues stabilized or destabilized the molecules against temperature. The combination of residues Asp, Cys, Glu, Lys, Leu, Arg, Trp, and Tyr which were polar-charged residues and nonpolar residues possessing high-surrounding hydrophobicity stabilized the molecules, while polar-uncharged residues Ala, Asp, Gly, Gln, Ser, Thr, Val, and Tyr destabilized them. A very high cooperativity exists among the stabilizing nonpolar residues, suggesting that their characteristic clustering inside the globule may enhance the thermostability of a protein. They concluded that the thermal stability of a protein could not be accounted for by a single type of force alone but by forces playing competing roles, e.g., ion-ion interaction, hydrophobic clustering etc.

Although still controversial, involvement of hydrophobic force in changes in protein molecules upon heating is probably undeniable. Koshiyama et al.[42] reported that an increase of hydrophobic region at high ionic strength indicated the possibility of stabilization of the quarternary structure of soy 11S globulin by hydrophobic bonding during heat denaturation.

Shimada and Matsushita[43] distinguished coagulation-type proteins (concentration dependent) and gelation-type proteins (concentration independent) based on the mole percent content of hydrophobic amino acid residues in the proteins (Figure 10). Concentration dependency seems to require two aspects of coagulable conditions. The protein should have a high molecular weight above 60,000 and a high content of hydrophobic groups.

Voutsinas et al.[7] used eight food proteins for analyzing thickening, heat coagulation, and gelation. The number of protein samples decreased to eight since some food proteins were so heat-stable that they did not yield measurable values for dependent variables in regression analysis. Heat coagulation (HC%) data only is shown here:

$$HC = -5.62 - 2.495SH + 0.009S_e \cdot SH \ (R^2 = 0.74, P < 0.05) \qquad (6)$$

where SH is the SH-group content ($\mu M/g$) and S_e is the exposed hydrophobicity as described before. The contribution of SH content is dependent on protein concentration in solutions. This is highly significant for thermal properties at high protein concentrations, e.g., thermally induced gelation, however, the contribution becomes negligible at lower protein concentra-

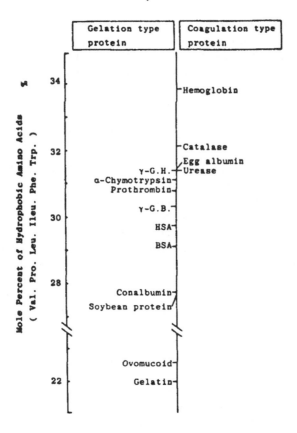

FIGURE 10.   Relationship between network structure type and
the mole percent of hydrophobic amino acids. The mole percent
of hydrophobic amino acids was calculated from the results of
amino acid analysis. HSA, human serum albumin; BSA, bovine
serum albumin; γ-G.H., γ-globulin human; γ-G.B., γ-globulin
bovine. (From Shimada, K. and Matsushita, S., *J. Agric. Food
Chem.*, 45, 2755, 1981. With permission.)

tions. S_e was more closely related to the thermal properties than S_o. An increase in S_o value
upon heating was reported by Kato et al.,[44] indicating unfolding of protein molecules, thereby
exposing hydrophobic sites located inside the molecules. According to Schmidt,[45] hydrophobic interactions are important in dissociative-associative reactions which initiate the
gelation process. These attractions could also be involved in layering or thickening of the
gel network strands upon cooling, which results in improved strength and stability.

V. WATER- AND FAT-BINDING CAPACITIES

Water-holding capacity is an important functionality for meat processing as it affects
tenderness, juiciness, color, taste, cookloss and drip on freezing and thawing. Recently,
Porteous and Wood[46] defined four parameters described as absorbed moisture, and retained
moisture after cooking, using distilled water (AM_w and RM_w) and 1% NaCl (AM_b and RM_b).
After mixing ground meat samples with water or brine, the equilibrated mixtures were
centrifuged and the gain in weight after decanting the unabsorbed water or brine was reported
as absorbed moisture. The values measured in the same manner after the samples were
heated at 78°C for 5 min were reported as retained moisture (Figure 11). The RM_b values
showed the highest correlation with the average exudate from vacuum-packaged wieners

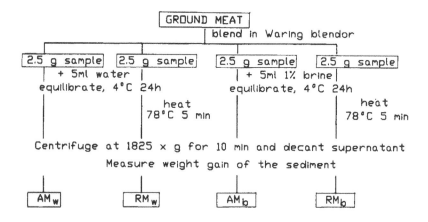

FIGURE 11.   Water-binding capacity of meat according to Porteous, J.D., and Wood, D.F.,[6] *Can. Inst. Food Sci. Technol. J.*, 16, 212, 1983.

after 7 and 14 days of refrigerated storage. Importance of protein hydrophobicity in these water-binding properties is discussed in Chapter 4.

Fat-binding capacity of proteins is also important as it enhances flavor retention and improves mouth feel. Fat-absorption of proteins is usually measured by adding excess liquid fat (oil) to a protein powder, thoroughly mixing and holding, centrifuging, and determining the amount of bound or adsorbed oil, which is total fat minus free fat.[34,47] The amount of oil and protein sample, kind of oil, holding, and centrifuging conditions, and units of expression have varied from one investigator to another.[48]

The mechanism of fat absorption is not clear. However, Wang and Kinsella[34] have attributed fat absorption, as assessed by the above methods, mostly to physical entrapment of the oil; in support of this a correlation coefficient of 0.95 was found between bulk density and fat absorption by alfalfa leaf proteins. Chemical modification of protein which increases bulk density concomitantly enhances fat absorption.[49]

Voutsinas and Nakai[9] proposed a method for measuring fat-binding capacity which reflects the true fat-binding capacity of proteins by minimizing the physical-entrapping effects. The method was later modified for applying to meat extracts, (Figure 12).[50] After centrifuging a mixture of dried protein and corn oil, the precipitate was suspended in $0.2N$ metaphosphoric acid. The metaphosphoric acid solution suspends the protein particles without dissolving them. After centrifuging the suspension and removing free oil, the precipitate was dispersed in the digestion medium ($7M$ urea in 50% H_2SO_4), then the turbidity due to oil bound to the protein was measured.

Factors affecting the protein-lipid interaction include protein conformation, protein-protein interactions, and the spatial arrangement of the lipid phase resulting from the lipid-lipid interaction.[48] Noncovalent interactions, e.g., hydrophobic, electrostatic, and hydrogen-bonding forces, are involved in the protein-lipid interactions. Hydrogen bonding is of secondary importance in lipid-protein complex, although it is indirectly important in hydrophobic bonding,[51] since in aqueous media the water-water interactions by hydrogen bonding are much stronger than the interaction between water and nonpolar groups, thus giving rise to hydrophobic bonding between nonpolar groups. Electrostatic attraction can occur between the negatively charged phosphate groups of phospholipids and positively charged protein groups, e.g., lysyl and guanidyl, or between a positively charged group in the phospholipid, e.g., choline, and a negatively charged amino acid side chain, e.g., glutamyl and aspartyl. A related mode of binding is the formation of salt bridges between a negatively charged phosphate group of a phospholipid via divalent calcium or other metal ions.[51-53] Hydrophobic

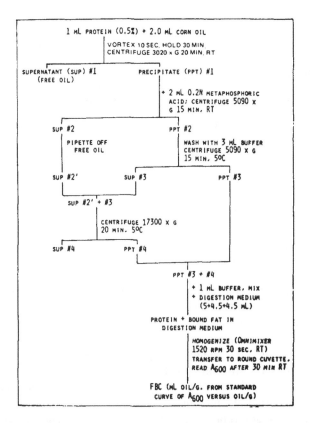

FIGURE 12.   Flow diagram of turbidimetric method for determination of fat-binding capacity (FBC) of muscle protein extracts. (From Li-Chan, E., et al., *J. Food Sci.*, 50, 1034, 1985. With permission.)

interaction is likely to play a major role in stabilizing the interactions of both polar and nonpolar lipids with proteins.[53]

Fat-binding capacity (FBC) was correlated with CPA hydrophobicity (S_o) and solubility (s) by Voutsinas and Nakai[9] as follows:

$$FBC = 30.27 + 1.38S_o - 0.014S_o \cdot s \ (R^2 = 0.902, \ n = 11, \ P < 0.01) \qquad (7)$$

VI. SUMMARY OF HYDROPHOBICITY-FUNCTIONALITY RELATIONSHIP

Relationships between structure and functionality of food proteins are schematically summarized in Figure 13. Solubility, hydrophobicity, charge, polymerization (dissociation or association), concentration, and sulfhydryl (SH)-disulfide (SS) content, all in complex combinations, affect the functionality, e.g., emulsification, water-holding, fat-binding, foaming, and gelation. It is likely that on each scale, especially hydrophobicity, an optimum location exists for the functionality in question as shown within the rectangle in Figure 13. Many more other functionalities, e.g., film formation, dough- and bread-making properties, texturization, etc. are important in food systems but literature on their correlation study using quantitative hydrophobicity scale values is virtually nonexistent.

VII. STATISTICAL DATA PROCESSING

The most well-known rule with regard to hydrophobicity or polarity is "like dissolves like". This means that there is always matched polarity between solvent and solute for the

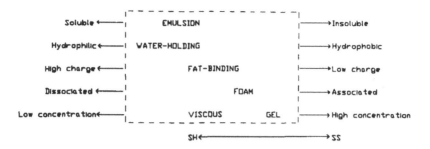

FIGURE 13. Structure-functionality relationship of food proteins.

best dissolution, implicating nonlinearity of the relationship. Although the polarity scale is designed to be linear, i.e., the polarity of a mixture of solvents can be additive, solubility of solute keeps increasing until the polarity of solvent matches, then decreases; therefore, the solubility-solvent polarity relationship is bell-shaped. A similar relationship exists between emulsifier and oil, with HLB and required HLB, which should be matched for the best emulsification of the oil.

A. Regression Analysis

The most logical expression of the above relationships can be obtained by polynomial curve fitting, usually using quadratic or cubic functions:

$$y = b_0 + b_1x + b_2x^2$$

$$\text{or} \quad y = b_0 + b_1x + b_2x^2 + b_3x^3 \tag{8}$$

For more than two predictor variables, multiple regression analysis has to be used. For three variables, e.g., solubility (x_1), hydrophobicity (x_2), and SH content (x_3) as functions for a functional property (y),

$$y = b_0 + b_1x_1 + b_2x_2 + b_3x_3 + b_{12}x_1x_2 + b_{13}x_1x_3 + b_{23}x_2x_3 + b_{11}x_1^2 +$$

$$b_{22}x_2^2 + b_{33}x_3^2 \tag{9}$$

which is a factorial form of quadratic equation more frequently used. The regression coefficient $\mathbf{b}$ can be computed by matrix algebra:

$$\mathbf{b} = (\mathbf{X'X})^{-1} \mathbf{X'Y} \tag{10}$$

and the square of the multiple correlation coefficient, or coefficient of multiple determination, R^2, is computed from:

$$R^2 = (\mathbf{b'X'Y} - n\bar{Y}^2)/(\mathbf{Y'Y} - n\bar{Y}^2) \tag{11}$$

Since R^2 is also $\Sigma(\hat{Y}_i - \bar{Y})^2/\Sigma(Y_i - \bar{Y})^2$, it represents the proportion of regression that accounts for the total deviation. The contributing proportion of a predictor variable against total variation in R^2 is computed as (regression coefficient) × (standard deviation) × (partial correlation coefficient) as described by Barylko-Pikielna and Metelski.[54]

Stepwise multiple regression analysis eliminates predictor variables one at a time in backward stepwise analysis, or adds one after the other in forward stepwise analysis, in order to find significant variables in the regression analysis without being perturbed with nonsignificant variables. It is difficult to obtain identical regression equations from forward vs. backward stepwise computations.[55] Schutz[56] recommended using levels of significance between 0.1 and 0.15 for the forward method, and 0.05 or 0.1 for the backwards method. In the SAS program,[57] an added variable is not always retained, being deleted at any step when it does not produce a significant F-statistic. Improved consistency was obtained by this method between the regression equations obtained from both ways of the stepwise computation.

If the predictor variables X are linearly correlated with each other (multicollinearity), $X'X$ does not have an inverse, or becomes numerically unstable, then the diagonal entries of $(X'X)^{-1}$ will be large and beyond a reasonable range. This yields a large variance for the coefficient b, implicating difficulty in obtaining significant regression coefficients.[58] The problems caused by collinearity can be alleviated or overcome by (1) deleting one of a pair of predictor variables that are strongly correlated, (2) use of ridge regression or (3) relating the response Y to the principal components of the predictor variables.

B. Ridge Regression

In the ridge regression analysis, the regression coefficients are computed as:

$$b = (X'X + k\ I)^{-1}\ X'Y \quad (k \geq 0) \tag{12}$$

instead of Equation 10. The ridge trace k is selected such that a stable set of regression coefficients is obtained.

For emulsion capacity of protein solutions, regular regression analysis yielded:

$$y = 54.991 - 3.301x_1 - 0.656x_2 + 0.404x_1^2 + 0.241x_2^2 + 0.012x_1x_2 \tag{13}$$

whereas by using ridge regression analysis the following equation was calculated:

$$y = 46.487 + 0.377x_1 - 0.098x_2 + 0.087x_1^2 + 0.071x_2^2 + 0.029x_1x_2 \tag{14}$$

where y is the emulsifying capacity, and x_1 and x_2 are pH and %NaCl, respectively.[59] Incorrect signs and too large regression coefficients are obvious as an effect of collinearity between pH and NaCl content in this case.

C. Principal Component Analysis

In principal component analysis, correlation coefficient matrix R is transferred to eigenvalue matrix Λ, that is a diagonal matrix without intercorrelations:

$$e'R = \Lambda\ e' \tag{15}$$

where e is eigenvector. Principal component score Y is then calculated by:

$$Y = Xe \tag{16}$$

where X is standard original data matrix.

The main purpose of a principal component analysis is to determine parameters, i.e., principal components, in order to explain as much of the total variation in the original data as possible with as few principal components as possible. The number of variables is reduced while maintaining as much of the original information as is possible by transforming the original set of variables into a smaller set of linear combinations in the form of principal components. The original data are finally expressed by a two dimensional illustration (canonical plot).

D. Classification

Classification is also a way to simplify the complex objects, i.e., a method to reduce the number of variables. There are two major techniques commonly being used in multivariate analysis: cluster analysis and discriminant analysis. The difference between them is the absence or presence of the groups to be classified which are known *a priori*.

1. Cluster Analysis

Cluster analysis is a technique to maximize the between-cluster variation relative to the within-cluster variation. Fundamental to the use of any clustering technique is the computation of similarity or, vice versa, distance between the respective objects. There are two types of scales to measure the similarity: *distance scale* and *matching scale*. The latter is used mainly for survey-type work with matched (1) or unmatched (0) responses for non-quantitative data.

Minkowski distance is defined as:

$$d_{ij} = \{\Sigma(x_{ik} - x_{jk})^r\}^{1/r} \tag{17}$$

When $r = 2$, it is the most popular Euclidean distance.

2. Discriminant Analysis

In order to maximize the between-group variance $(\hat{b}'Bb)$ relative to the within-group variance $(b'Wb)$, the following relationship exists:

$$(W^{-1}B - \hat{\lambda}I)\,\hat{b} = 0 \tag{18}$$

where W and B are the matrix within-group and between-group sums of squares and cross products, respectively. This equation is solved for eigenvalue $\hat{\lambda}$ and corresponding eigenvector $\hat{b}$ of the matrix $W^{-1}B$. Eigenvectors corresponding to the largest and second-largest eigenvalues are used as a set of discriminant weights in two linear discriminant equations. The separation of group-means and the scattergram of individual data points are demonstrated in the two-dimensional discriminant space.

REFERENCES

1. **Wolf, W. J.,** Soy proteins: their functional, chemical, and physical properties, *J. Agric. Food Chem.,* 18, 969, 1970.
2. **Kinsella, J. E.,** Functional properties of proteins in foods: a survey, *Crit. Rev. Food Sci. Nutr.,* 7, 219, 1976.
3. **Hermansson, A.-M.,** Functional properties of proteins for food stability, *AULs halvarsskrift,* 2, 1973.
4. **Aoki, H., Taneyama, O., and Inami, M.,** Emulsifying properties of soy protein: characteristics of 7S and 11S proteins, *J. Food Sci.,* 45, 534, 1980.
5. **McWatters, K. H. and Holmes, M. R.,** Influence of moist heat on solubility and emulsification properties of soy and peanut flours, *J. Food Sci.,* 44, 774, 1979.
6. **Voutsinas, L. P., Cheung, E., and Nakai, S.,** Relationships of hydrophobicity to emulsifying properties of heat denatured proteins, *J. Food Sci.,* 48, 26, 1983.
7. **Voutsinas, L. P., Nakai, S., Harwalkar, V. R.,** Relationships between protein hydrophobicity and thermal functional properties of food proteins, *Can. Inst. Food Sci. Technol. J.,* 16, 185, 1983.
8. **Townsend, A.-A. and Nakai, S.,** Relationships between hydrophobicity and foaming characteristics of food proteins, *J. Food Sci.,* 48, 588, 1983.

9. **Voutsinas, L. P. and Nakai, S.,** A simple turbidimetric method for determining the fat binding capacity of proteins, *J. Agric. Food Chem.,* 31, 58, 1983.

10. **Li-Chan, E., Nakai, S., and Wood, D. F.,** Hydrophobicity and solubility of meat proteins and their relationship to emulsifying properties, *J. Food Sci.,* 49, 345, 1984.

11. **Bigelow, C. C.,** On the average hydrophobicity of proteins and the relation between it and protein structure, *J. Theoret. Biol.,* 16, 187, 1967.

12. **Hayakawa, S. and Nakai, S.,** Relationships of hydrophobicity and net charge to the solubility of milk and soy proteins, *J. Food Sci.,* 50, 486, 1985.

13. **Burley, R. W. and Sleigh, R. W.,** Hydrophobic chromatography of proteins in urea solutions. The separation of apoproteins from a lipoprotein of avian egg yolk, *Biochem. J.,* 209, 143, 1983.

14. **Porath, J.,** Gel filtration of proteins, peptides and amino acids, *Biochim. Biophys. Acta,* 39, 193, 1960.

15. **Hofstee, B. H. J. and Otillio, F. N.,** Modifying factors in hydrophobic protein binding by substituted agaroses, *J. Chromatog.,* 161, 153, 1978.

16. **Snyder, L. R.,** Classification of the solvent properties of common liquids, *J. Chromatog. Sci.,* 16, 223, 1978.

17. **Kamlet, M. J., Abboud, J. L., and Taft, R. W.,** The solvatochromic comparison method, 6. The π^* scale of solvent properties, *J. Am. Chem. Soc.,* 99, 6027, 1977.

18. **Suzuki, K. and Imahori, K.,** Glyceraldehyde 3-phosphate dehydrogenase of *Bacillus stearothermophilus, J. Biochem.,* 74, 955, 1973.

19. **Hachimori, A., Shiroya, Y., Hirato, A., Miyahara, T., and Samejima, T.,** Effects of divalent cations on thermophilic inorganic pyrophosphatase, *J. Biochem.,* 86, 121, 1979.

20. **Mozhaev, V. V. and Martinek, K.,** Structure-stability relationships in proteins: new approaches to stabilizing enzymes, *Enzyme Microb. Technol.,* 6, 50, 1984.

21. **Burley, S. K. and Petsko, G. A.,** Aromatic-aromatic interaction: a mechanism of protein structure stabilization, *Science,* 229, (4708), 23, 1985.

22. **Kato, A. and Nakai, S.,** Hydrophobicity determined by a fluorescence probe method and its correlation with surface properties of proteins, *Biochim. Biophys. Acta,* 624, 13, 1980.

23. **Nakai, S., Ho, L., Helbig, N., Kato, A., and Tung, M. A.,** Relationship between hydrophobicity and emulsifying properties of some plant proteins, *Can. Inst. Food Sci. Technol. J.,* 13, 23, 1980.

24. **Pearce, K. N. and Kinsella, J. E.,** Emulsifying properties of proteins: evaluation of a turbidimetric technique, *J. Agric. Food Chem.,* 26, 716, 1978.

25. **Aoki, H., Taneyama, O., Orimo, N., and Kitagawa, I.,** Effect of lipophilization of soy protein on its emulsion stabilizing properties, *J. Food Sci.,* 46, 1192, 1981.

26. **Cherry, J. P. and McWatters, K. H.,** Whippability and aeration, in *Protein Functionality in Foods,* Cherry, J. P., Ed., ACS Symp. Ser. 147, American Chemical Society, Washington, D.C.

27. **German, J. B., O'Neill, T., and Kinsella, J. E.,** Film forming and foaming behavior of food proteins, *J. Am. Oil Chem. Soc.,* 62, 1358, 1985.

28. **Kato, A., Fujimoto, K., Matsudomi, N., and Kobayashi, K.,** Protein flexibility and functional properties of heat-denatured ovalbumin and lysozyme, *Agric. Biol. Chem.,* 50, 417, 1986.

29. **Graham, D. E. and Phillips, M. C.,** The conformation of proteins at the air-water interface and their role in stabilizing foams, in *Foams,* Akers, R. J., Ed., Academic Press, New York, 1976, 15.

30. **Phillips, M. C.,** Protein conformation at liquid interfaces and its role in stabilizing emulsions and foams, *Food Technol.,* 35, 50, 1981.

31. **Kim, S. H. and Kinsella, J. E.,** Surface activity of food proteins: relationship between surface pressure development, viscosity of interfacial films and foam stability of bovine albumin, *J. Food Sci.,* 50, 1526, 1985.

32. **Horiuchi, T., Fukushima, D., Sugimoto, H., and Hattori, T.,** Studies on enzyme-modified proteins as foaming agents: effect of structure on foam stability, *Food Chem.,* 3, 35, 1978.

33. **Hermansson, A.-M., Sivik, B., and Skjoldebrand, C.,** Functional properties of protein for food — factors affecting solubility, foaming and swelling of fish protein concentrate, *Lebensm.- Wiss. u. -Technol.,* 4, 201, 1971.

34. **Wang, J. C. and Kinsella, J. E.,** Functional properties of alfalfa leaf protein: foaming, *J. Food Sci.,* 41, 498, 1976.

35. **Waniska, R. D. and Kinsella, J. E.,** Foaming properties of proteins: evaluation of a column aeration apparatus using ovalbumin, *J. Food Sci.,* 44, 1398, 1979.

36. **Graham, D. E. and Phillips, M. C.,** Proteins at liquid interfaces. III. Molecular structures of adsorbed films, *J. Colloid Interface Sci.,* 70, 427, 1979.

37. **Brandts, J. F.,** Heat effects on proteins and enzymes, in *Thermobiology,* Rose, A. H., Ed., Academic Press, New York, 1967, 3.

38. **Singleton, R., Jr. and Amelunxen, R. E.,** Proteins from thermophilic microorganisms, *Bacteriol. Rev.,* 37, 320, 1973.

39. **Goldsack, D. E.,** Relation of the hydrophobicity index to the thermal stability of homologous proteins, *Biopolymers,* 9, 247, 1970.

40. **Ikai, A.,** Thermostability and aliphatic index of globular proteins, *J. Biochem.,* 88, 1895, 1980.

41. **Ponnuswamy, P. K., Muthusamy, R., and Manavalan, P.,** Amino acid composition and thermal stability of proteins, *Int. J. Biol. Macromol.,* 4, 186, 1982.

42. **Koshiyama, I., Hamano, M., and Fukushima, D.,** A heat denaturation study of the 11S globulin in soybean seeds, *Food Chem.,* 6, 309, 1980-81.

43. **Shimada, K. and Matsushita, S.,** Relationship between thermocoagulation of proteins and amino acid compositions, *J. Agric. Food Chem.,* 28, 413, 1980.

44. **Kato, A., Tsutsui, N., Kobayashi, K., and Nakai, S.,** Effects of partial denaturation on surface properties of ovalbumin and lysozyme, *Agric. Biol. Chem.,* 45, 2755, 1981.

45. **Schmidt, R. H.,** Gelation and coagulation, in *Protein Functionality in Foods,* Cherry, J. P., Ed., ACS Symp. Ser. 147, American Chemical Society, Washington, D.C., 1981, 7.

46. **Porteous, J. D. and Wood, D. F.,** Water-binding of red meats in sausage formation, *Can. Inst. Food Sci. Technol. J.,* 16, 212, 1983.

47. **Lin, M. J., Humberts, E. S., and Sosulski, F. W.,** Certain functional properties of sunflower meal products, *J. Food Sci.,* 39, 368, 1974.

48. **Hutton, C. W. and Campbell, A. M.,** Water and fat absorption, in *Protein Functionality in Foods,* Cherry, J. P., Ed., ACS Symp. Ser. 147, American Chemical Society, Washington, D.C., 1981, 9.

49. **Franzen, K.,** Ph.D. thesis, Cornell University, 1975, cited by Kinsella, J. E., *CRC Crit. Rev. Food Sci. Nutr.,* 7, 219, 1976.

50. **Li-Chan, E., Nakai, S., and Wood, D. F.,** Relationship between functional (fat binding, emulsifying) and physicochemical properties of muscle proteins. Effects of heating, freezing, pH and species, *J. Food Sci.,* 50, 1034, 1985.

51. **Karel, M.,** Protein-lipid interactions, *J. Food Sci.,* 38, 756, 1973.

52. **Pomeranz, Y.,** Interaction between glycolipids and wheat flour macromolecules in breadmaking, *Adv. Food Res.,* 20, 153, 1973.

53. **Ryan, D. S.,** Determinants of the functional properties of proteins and protein derivatives in foods, in *Food Proteins: Improvement through Chemical and Enzymatic Modifications,* Feeney, R. E., and Whitaker, J. R., Eds., ACS Adv. Chem. Ser. 160, American Chemical Society, Washington, D.C., 1977, 4.

54. **Barylko-Pikielna, N. and Metelski, K.,** Determination of contribution coefficients in sensory scoring of over-all quality, *J. Food Sci.,* 29, 109, 1964.

55. **Hocking, R. R.,** The analysis and selection of variables in linear regression, *Biometrics,* 32, 1, 1976.

56. **Schutz, H. G.,** Multiple regression approach to optimization, *Food Technol.,* 37(11), 46, 1983.

57. *SAS User's Guide: Statistics,* SAS Institute, Cary, N.C., 1985.

58. **Johnson, R. A. and Wilchern, D. W.,** *Applied Multivariate Statistical Analysis,* Prentice-Hall, Englewood Cliffs, N.J., 1982.

59. **Newell, G. J. and Lee, B.,** Ridge regression: an alternative to multiple linear regression for highly correlated data, *J. Food Sci.,* 46, 968, 1981.

Chapter 4

HYDROPHOBIC INTERACTIONS IN MUSCLE PROTEINS AND THEIR ROLE IN COMMINUTED MEAT SYSTEMS

TABLE OF CONTENTS

I. INTRODUCTION

The term "meat" generally refers to muscle which has undergone certain postmortem chemical and biochemical changes prior to consumption. It includes the muscle foods derived from various species of animals, including poultry meat and the red meats such as beef, pork, veal, and mutton. Not generally included in the term "meat" are the muscle foods derived from nonmammalian species such as fish, although "fish meat" is sometimes used for this category of muscle food. In this chapter, the terms "muscle" and "meat" are used somewhat interchangeably, with the forewarning by Hultin[1] that generally speaking, "muscle" is probably the more appropriate term for the functional tissue and "meat" is the more appropriate term for the tissue after it has passed through certain changes after slaughter of the animal.

The lean portion of meat contains roughly 20% protein, with only slight variations in the content between different species (Table 1). In addition to their high content in muscle tissues, the muscle proteins are valued for their high nutritional quality and superior functional performance. Thus, various meats are valued for consumption not only in the form of the intact muscle tissue, but also in comminuted meat products such as sausages as well as restructured meat products.

Muscle protein biochemistry and chemistry are ever-expanding fields of research. The past decade has seen great strides in the advancement of muscle protein research, including studies on the molecular basis for differences between muscle fiber activity,[2,3] discovery of new minor regulatory and structural proteins,[4] and degradation and appearance of new components in the muscle ultrastructure during postmortem aging treatments.[5] The molecular mechanism underlying muscle contraction and relaxation processes, their control, and accompanying biochemical changes, are topics covering an extensive field which has been, and still is, the center of intensive basic research. These studies show great potential for application towards improving the eating quality of intact muscles, including meat tenderness.[5]

The great variability frequently observed in muscle properties appears to arise from a complex interaction of intrinsic factors as well as extrinsic factors. In the latter category are pre and postslaughter conditions, and processing and postmortem storage conditions. In the former category are parameters such as species, breed, sex, age, anatomical location of muscle, training or exercise, nutrition, and interanimal variability.[6] The least understood of these intrinsic factors appears to be variability between individual animals; even between littermates of the same sex, considerable differences have been noted in percentages of intramuscular fat, moisture, and total nitrogen, as well as relative distribution of nitrogen between the sarcoplasmic, myofibrillar, and stroma proteins.

Despite the abundance of literature on muscle proteins, there is, in fact, a scarcity of information which a meat processor could apply directly to control the quality of processed meat products in the presence of high variability of ingredients. Currently, the meat-processing industry uses linear programming for ingredient-blending formulation, with the objective of obtaining a formula of the minimum cost, subject to constraints of fat, moisture, and protein content. However, the quality of comminuted or restructured meat products is dependent on the functionality of the protein matrix within the products. In attempts to maintain constant quality, various "bind" constants based on emulsifying capacity, emulsifying stability, soluble protein, and/or total protein content of the ingredients have been suggested.[7-9] However, the poor predictability of solubility and emulsifying capacity as an indicator of the product functionality, and the question of whether or not a comminuted meat system can be considered a real emulsion,[10,11] have cast increasing doubt on the reliability of the bind-constant system. Comer and Dempster[12] suggested an alternative approach based on protein quality factors or a functional property test to reflect gelation properties of ingredients. However, the basis for protein quality factors was rather empirical,

Table 1
PROXIMATE COMPOSITION OF
LEAN MUSCLE TISSUE FROM
DIFFERENT SPECIES

Species	Composition (%)			
	Water	Protein	Lipid	Ash
Beef	70—73	20—22	4—8	1
Pork	68—70	19—20	9—11	1.4
Chicken	74	20—23	5	1
Lamb	73	20	5—6	1.6
Cod	81	17—18	0.3	1.2
Salmon	64	20—22	13—15	1.3

Adapted from Hultin, H. O., in *Principles of Food Science, Part 1. Food Chemistry*, Fennema, O. R., Ed., Marcel Dekker, New York, 1976, 577.

being subjectively chosen. There is thus a need for objective and reliable methods to evaluate the functionality of these ingredients. An alternative approach is to apply the quantitative structure-activity relationship to predict functionality of the ingredients from basic properties of the proteins, including properties such as charge, surface hydrophobicity, and solubility.

The following discussion presents an overview of current knowledge on the characteristics of muscle protein fractions which are responsible for their functional properties. Rather than focussing on the intact muscle tissue properties, the properties of muscle proteins important for their use as ingredients in comminuted meat products or restructured products are emphasized. The application of a quantitative structure-activity relationship (QSAR) approach to predict functional properties of muscle proteins from measurements of their physico-chemical properties, is described.

II. MUSCLE PROTEIN COMPOSITION

Traditionally, the proteins found in muscle have been classified into categories according to their distribution, organization, solubility, and function in the living muscle tissue. Thus, intracellular and extracellular proteins are defined as those proteins located within and outside of the sarcolemmal membrane of muscle syncytia, respectively.[4] The muscle proteins are also frequently classified on the basis of solubility. The three broad solubility classes are (1) proteins which are soluble in water or dilute salt solution (sarcoplasmic fraction), (2) proteins which are soluble in more concentrated salt solution (myofibrillar fraction), and (3) proteins which are insoluble even in solutions of high salt concentration (connective tissue or stromal protein fraction). Table 2 lists the major proteins within these categories for typical adult mammalian muscle. Ongoing extensive research by many groups throughout the world, necessitates continual updating to provide a complete list of known muscle protein components. A recent reveiw by Ashgar et al.,[4] for example, cites that about 100 different proteins are known to be in the sarcoplasmic fraction. While the significance fo these numerous proteins to the biological function of living muscle tissue cannot be discounted, it is beyond the scope of this text to elaborate on these proteins, which may be considered minor, at least in quantitative aspects.

A. Sarcoplasmic Proteins

The sarcoplasmic proteins can be extracted from muscle with water or dilute salt solutions (ionic strength <0.1). This fraction constitutes about 30 to 35% of the total muscle protein,

Table 2
CHEMICAL COMPOSITION OF TYPICAL ADULT MAMMALIAN POSTRIGOR MUSCLE

Components	% Wet weight
Water	75.0
Protein	19.0
Myofibrillar	11.5
Myosin[a]	5.5
Actin[a]	2.5
Connectin	0.7
Tropomyosins	0.6
Troponins (TnI, TnC, TnT)	0.6
α-, β- and γ-actinins	0.5
Myomesin, N-line, and C-proteins	0.7
Desmin, filamin, F- and I-proteins, etc.	0.4
Sarcoplasmic	5.5
Glyceraldehyde phosphate dehydrogenase	1.2
Aldolase	0.6
Creatine kinase	0.5
Other glycolytic enzymes	2.2
Myoglobin	0.2
Haemoglobin and other unspecified extracellular proteins	0.6
Connective tissue and organelle	2.0
Collagen	1.0
Elastin	0.05
Mitochondrial, etc., (including cytochrome c and insoluble enzymes)	0.95
Lipid	2.5
Neutral lipid, phospholipids, fatty acids, fat-soluble substances	
Carbohydrate	1.2
Lactic acid, glucose-6-phosphage, glycogen, glucose, traces of other glycolytic intermediates	
Miscellaneous soluble nonprotein substances	2.3
Nitrogenous	1.65
Creatinine, inosine monophosphate, di- and triphosphopyridine nucleotides, amino acids, carnosine, anserine	
Inorganic	0.65
Total soluble phosphorus, potassium, sodium, magnesium, calcium, zinc, trace metals	
Vitamins	
Various fat- and water-soluble vitamins, quantitatively minute	

[a] Actin and myosin are combined as actomyosin in postrigor muscle.

Adapted from Lawrie, R. A., in *Meat Science*, 4th ed., Lawrie, R. A., Ed., Pergamon Press, New York, 1985.

or roughly 5% of the wet weight of the muscle. About 100 different proteins are known to be present in the sarcoplasmic fraction, including the enzymes of the glycolytic pathway and the muscle pigment myoglobin.[4] Although diverse in their biological functions, the proteins in this fraction appear to share many common general physicochemical characteristics, including their relatively low molecular weight, globular or rod-shaped conformation, and low viscosity. They do, however, differ in various properties such as charge (which allows their separation and identification by electrophoretic techniques) and susceptibility to denaturation.[6] It is also interesting to note that while these proteins are extracted at low ionic strength in vitro, many of them, including the glycolytic enzymes, are bound to the myofibrillar protein, actin, in vivo, which is probably related to their control of enzymic reactions in the muscle.[13] These proteins are coextracted with the myofibrillar proteins from muscle using solutions with high concentrations of salt, including the 2.5% sodium chloride levels commonly used for comminuted meat products.

B. Myofibrillar Proteins

The myofibrillar proteins have in common the property of requiring salt solutions of high ionic strength for extraction from muscle. Once extracted, however, many of these proteins are soluble in solutions of lower ionic strength. The proteins in this fraction make up the myofibrils within the muscle fibers, and may be further classified into the contractile, regulatory, and structural proteins, based on their physiological function. Together, these myofibrillar proteins constitute 55 to 60% of the total muscle protein or roughly 10% of the wet weight of vertebrate skeletal muscle. The major contractile proteins are myosin and actin, which constitute the main components of the thick and thin filaments, respectively. These proteins are directly involved in the contraction and relaxation of living muscle, and are also the dominant proteins in muscle in terms of composition. Myosin comprises 43%, while actin comprises 22% of the myofibrillar fraction.[14]

Regulatory proteins are those myofibrillar proteins which, though not directly involved in cross-bridge formation, have important functions in regulating the processes of contraction and relaxation. The major regulatory proteins, tropomyosin and troponin, which are located in the thin filaments, impart calcium sensitivity to the actomyosin system. The tropomyosin molecule consists of two polypeptides called α- and β-chain which have a helical conformation, with 3.5 residues per turn. The two chains coil around each other to form a superhelix with about nine residues overlapping on consecutive molecules. Extensive studies based on the tropomyosin sequence indicate a hydrophobic line of interaction when the chains are coiled in an α-helix with 3.5 or 3.6 residues per turn, and it has been postulated that the superhelix has a highly hydrophobic core, while polar residues are on the surface.[15] Each molecule of tropomyosin is believed to bind seven actin molecules.

The troponins consist of three components, designated the calcium-binding subunit (Troponin-C or TnC), the actomyosin ATPase inhibitory subunit (Troponin-I or TnI), and the tropomyosin-binding subunit (Troponin-T or TnT).

In addition to tropomyosin and troponin, other minor regulatory proteins are also present in the myofibrillar fraction, including the M-protein and creatine kinase located at the M-band, the actinins in the Z-band, and C-protein in the thick filament. While important for the control of muscle function in living muscle as well as in the chemical and biochemical changes occurring after slaughter of the animal, these proteins are relatively minor in quantitative terms. The reader interested in more detailed description of these proteins may refer to the recent review by Ashgar et al.[4] which includes a summary of properties of both the major and minor regulatory proteins (Table 3).

In addition to the contractile and regulatory proteins, another group of proteins is found in filaments which have been variously referred to as "S-filaments", "T-filaments", "gap filaments", and "intermediate filaments"[4,6,13] Morphologically, these filaments resemble collagen fibrils, but rather than being located extracellularly as in the latter case, they are situated within the muscle cells and are believed to strengthen the myofibrillar structure. Hence, they have also been named "scaffold" or "backbone" proteins. Included in this category of proteins are connectin (titin), nebulin, desmin (skeletin), vimentin, and synemin (Table 4).

Table 3
COMPOSITION OF REGULATORY PROTEINS OF VERTEBRATE SKELETAL MUSCLE

Protein	Location in ultrastructure	% of total protein	Molecular weight	Function
Tropomyosin	I-Band	5.0	65,000—70,000	Binds to actin and troponin
α-Chain			34,000	
β-Chain			36,000	
Troponins	I-Band	5.0		Ca^{++} regulation
Troponin-C			21,000	
Troponin-I			24,000[a]	Inhibits actomyosin ATPase, binds Ca^{++}
Troponin-T			35,000	Binds to tropomyosin
M-Protein	M-Filament	2.0	165,000	Binds to myosin
Creatine kinase	M-Bridges		42,000	
β-Protein	M-Line(?)		130,000	
C-Protein	Myosin filament	2.0	140,000	Binds to myosin
F-Protein	A-Band	0.1	121,000	Binds to myosin
I-Protein	A-Band	0.1	50,000	Binds to myosin
H-Protein	A-Band		69,000	Binds to myosin
X-Protein	A-Band		152,000	Binds to myosin
Paratropomyosin	Edges of A-Band		35,000	Possible role in postmortem changes
Actinins				
α-Actinin	Z-Line	2.0	95,000	Binds to actin (gelation of F-actin)
β-Actinin	Free ends of I-filament	0.1	37,000[b] and 34,000	Binds to actin
γ-Actinin	I-Band(?)	0.1	35,000	Binds to actin
Eu-Actinin	Z-line	0.1	42,000	Binds to actin
Z-Protein	Lattice structure of Z-line	0.3	55,000	
Z-Nin	Z-Line	0.4	300,000	

Note: (?) = not established.

[a] From Hofmann, K. and Hamm, R., *Adv. Food Res.*, 24, 2, 1978.
[b] Molecular weight of two subunits.

Table adapted from Ashgar, A. et al., *CRC Crit. Rev. Food Sci. Nutr.*, 22, 27, 1986.

C. Connective Tissue or Stroma Proteins

The third category of muscle proteins classified by solubility properties is the so-called "insoluble" fraction, which refers to the fraction remaining after extraction with water, dilute salt, and concentrated salt solutions. Collectively, it is often referred to as "connective tissue" or "stroma proteins" fraction, and it constitutes about 10 to 15% of the total muscle proteins or about 2% of the wet weight of the muscle; these proteins include collagen, reticulin, and elastin.

III. PROPERTIES OF MYOFIBRILLAR PROTEINS, PARTICULARLY MYOSIN AND ACTIN

Quantitatively, the two myofibrillar proteins myosin and actin are the most significant in muscle protein composition. At the same time, the overall functional properties significant in terms of the overall quality of meat and meat products have also been generally ascribed

Table 4
COMPOSITION OF SCAFFOLD PROTEINS OF VERTEBRATE SKELETAL MUSCLE

Protein	Location in structure	% of total protein	Molecular weight	Function
Connectin	A-I junction	5.0	>1,000,000	Nets of 2-nm filaments, binds to myosin and actin
Nebulin	N-line	5.0	500,000—600,000	—
Titin	A-I junction, Z-Line and H-zone	10.0	>1,000,000	Similar functions to connectin
Filamin	Z-line and I-Band	—	230,000	Gelation of F-actin
Desmin	Periphery of Z-line	1.0	550,000	10-nm filament linking neighboring Z-line
Synemin	Periphery of Z-line	1.0	220,000	Association with desmin and vimentin
Vimentin	Periphery of Z-line	1.0	58,000	Forms transverse structure

Adapted from Ashgar, A. et al. *CRC Crit. Rev. Food Sci. Nutr.*, 22, 27, 1986.

to myosin, which is the predominant protein in prerigor meat, or to its complex with actin known as actomyosin, which is the predominant form in postrigor meat. Thus for example, Goll et al.[16] estimated that approximately 97% of the water-holding capacity and over 75% of the emulsifying capacity of meat, are due to the myofibrillar proteins alone. Although tenderness may be more greatly influenced by connective tissue content, the myofibrillar fraction is still crucial in influencing tenderness, especially due to the great changes which may occur in the myofibrillar proteins upon postmortem storage.

Thus, although the other components of the myofibrillar fraction as well as the sarcoplasmic and connective fractons have some effects on the final quality, the majority of available evidence points to myosin and actomyosin as the major determinants of muscle protein functionality. Hence, in order to elucidate the basis for muscle protein functionality, it is necessary to have some basic understanding of the structure and molecular forces which are significant in these two proteins, and to investigate the changes in these properties upon slaughter and processing of muscle tissue.

A. Myosin

Myosin is a large, asymmetric protein roughly 160 nm in length having a rod-like shape with a globular tip.[17] Each myosin molecule generally consists of six subunits: two large polypeptide units, commonly referred to as "heavy chains", and four small polypeptide units commonly referred to as "light chains" (Figure 1). Each of the heavy chains has a molecular weight of about 200,000, while the light chains vary in molecular weight from about 16,000 to 25,000, so that each myosin molecule has a molecular weight of roughly 480,000 to 500,000.[4]

The myosin molecule is often described as possessing two globular "heads" attached to a rod-like "tail". In fact, the heavy chains each have a predominantly α-helix conformation and coil around each other for a great part of their length, forming a superhelix which constitutes the "tail". At the amino terminals of the two polypeptide chains, however, each heavy chain folds separately to form a globular structure or head. Each head of the myosin molecule contains one actin-binding site and one ATP-binding site, and possesses ATPase activity.

The light chains are located primarily in the head region. One of the light chains, designated "LC-1", is often named the "alkali light chain-1", referring to its isolation by alkali treatment. Its molecular weight usually falls in the range of 20,700 to 25,000. A second

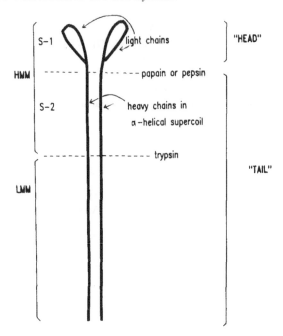

FIGURE 1.   Schematic diagram of the myosin molecule. Fragments obtained by enzymatic cleavage are HMM (heavy meromyosin) S – 1 and HMM S – 2, and LMM (light meromyosin). The myosin "head" refers to HMM S – 1, whereas the "tail" refers to LMM + HMM S – 2.

type of light chain, designated "LC-2", is also referred to as the "DTNB light chain" on the basis of its isolation with 5,5′-dithiobis(2-nitrobenzoic acid) or DTNB. Its molecular weight is approximately 19,000 to 20,000. The third type of light chain, designated "LC-3", is also isolated with alkali and may be referred to sometimes as "alkali light chain-2". Its molecular weight is 16,000 to 16,500. Generally, 4 light chains are found in each myosin molecule: one DTNB light chain and one alkali light chain-1 associate with one of the heavy chain heads, while a second DTNB light chain and one alkali light chain-2 associate with the other heavy chain head. The asymmetry of the two heads has led to their designation as "S-1A" for the heavier head and "S-1B" for the lighter head.

From studies on vulnerability to proteolytic enzyme action, the myosin molecule has been further classified into a light meromyosin (LMM) section on the tail, and a heavy meromyosin (HMM) section consisting of part of the tail and the head, with molecular weights of 150,000 and 350,000, respectively (Figure 1). A "hinge" region in the HMM is susceptible to further proteolysis, with the cleavage leading to a subfragment-1 (HMM-S1) containing the globular head, and a subfragment 2 (HMM-S2), containing the rod-like section of HMM. In fact, the hinge region which can be located in vitro by the site of proteolytic enzyme action is also a site of rotation in the intact myosin molecule under physiological conditions and it has been proposed that flexibility of this region allows the myosin head to project out during the genesis of filaments.

The myosins from normal muscles of different vertebrate species such as tuna, chicken, and rabbit, have markedly similar ATPase enzymic characteristics and seem to differ only slightly in amino acid composition (Table 5).[18] The importance of differences in amino acid composition for different myosins is difficult to assess because of their large molecular weights; potential errors thus preclude any final conclusion as to whether their amino acid compositions are in fact truly identical.[19] On the other hand, there is ample experimental evidence about the existence of polymorphism in myosin molecules. Myosins isolated from

Table 5

AMINO ACID COMPOSITIONS OF SKELETAL MUSCLE MYOSINS AND
MYOSIN FRAGMENTS

Amino acid	Myosin			H-Meromyosin	L-Meromyosin	L-Meromyosin Fraction I
	Beef	Pork	Rabbit			
Cystine/2[a]			9	11	6	
Aspartic acid	87	80	85	88	77	82
Threonine	43	47	41	49	38	36
Serine	34	37	41	43	37	38
Glutamic acid	146	150	155	138	174	204
Proline	26	20	22	29	8	2
Glycine	38	39	39	45	24	18
Alanine	73	65	78	73	76	84
Valine	37	39	42	45	39	28
Methionine	20	21	22	19	14	18
Isoleucine	40	41	42	42	35	32
Leucine	86	86	79	78	85	96
Tyrosine	16	16	18	21	12	9
Phenylalanine	27	28	27	40	10	4
Lysine	107	105	85	82	83	86
Histidine	14	15	15	12	19	22
Arginine	42	46	41	29	51	61
Tryptophan			4	3	6	
(NH₃)[b]	(106)	(120)	(86)	(100)	(107)	(118)
Total residues	(836)	(837)	845	847	794	(820)

Note: Residues per 10^5 g. All values rounded to nearest integer.

[a] Reported values for various derivatives of cysteine have been converted to cystine/2 equivalents.
[b] Figures in parentheses indicate lack of confidence in data.

Adapted from Bodwell, C. E. and McClain, P. E., in *The Science of Meat and Meat Products*, Price, J. F. and Schweigert, B. S., Eds., W. H. Freeman, San Francisco, 1971, 78.

white (slow) muscles and cardiac muscles have different enzymic properties than those from red (fast) muscles. The molecular basis of myosin isozymes differs from other types of isozymes, such as lactic dehydrogenase. In the latter case, the isozymes arise from different distribution of the same subunits, while in the case of myosin, the different isozymic forms represent heteropolymers of subunits which are characteristic of each muscle fiber type. Thus it appears that different sets of genes are active in producing components of myosin that make up the isozymic forms characteristic of each muscle type.[20] White muscle myosin has been shown to contain 3 light chain components designated: LC_{1W}, LC_{2W}, and LC_{3W}, while cardiac and red muscle myosin contain only two light chain components, designated LC_{1CR} and $LC2_{2CR}$; red muscle myosin also has an additional light chain LC_{1R} which is unique to it.[20,21] Immunochemical studies of myosins from chicken cardiac, fast white, slow red, and smooth muscle suggested that the heavy chains of myosins show very strict immunochemical specificity, each having their own specific antigenicities.[22] The slow muscle light chains and cardiac muscle light chains were found to have similar immunological characteristics while they differed immunologically from fast muscle light chains.[2]

These immunologically, biochemically, and physiologically significant differences between muscle fiber types may also be important to the functional properties of meat products. For example, salt-soluble proteins from turkey leg muscle ("dark" or red meat) had greater gelation capacity than those from breast muscle ("white" or light meat),[23] while chicken

Table 6
AMINO ACID COMPOSITIONS OF SKELETAL MUSCLE ACTINS AND ACTININS

Amino acid	Actin					Actinin	
	Beef	Pork	Lamb	Chicken	Rabbit	α	β
Cystine/2[a]	8	11	9	10	11	—	—
Aspartic acid	82	81	80	80	81	80	80
Threonine	69	67	67	64	61	54	54
Serine	59	58	58	56	53	43	46
Glutamic acid	93	95	96	93	95	102	89
Proline	41	43	42	44	47	49	46
Glycine	67	64	64	64	64	64	61
Alanine	72	68	70	67	68	67	67
Valine	44	40	43	42	38	40	42
Methionine	23	33	27	32	36	45	33
Isoleucine	63	63	63	63	48	58	63
Leucine	62	61	63	62	63	66	65
Tyrosine	32	34	35	36	37	35	34
Phenylalanine	28	28	27	28	28	29	27
Lysine	46	44	47	44	48	43	52
Histidine	18	17	18	17	18	19	19
Arginine	42	41	43	40	45	41	36
Tryptophan	8	9	6	8	10		
(NH₃)[b]	—	—	—	—	—	(66)	(75)
Total residues	857	857	857	850	837	835	814

Note: Residues per 10^5 g. All values rounded to nearest integer.

[a] Values for derivatives of cysteine are expressed as cystine/2.
[b] Figures in parentheses indicate lack of confidence in data.

Adapted from Bodwell, C. E. and McClain, P. E., in *The Science of Meat and Meat Products*, Price, J. F. and Schweigert, B. S., Eds., W. H. Freeman, San Francisco, 1971, 78.

white muscle myosin was demonstrated to consistently yield stronger gels than red muscle myosin gels.[3,24]

B. Actin

Actin may exist in either a monomeric globular form (G-actin) or a polymeric fibrous form (F-actin), depending on environmental conditions. The monomeric G-actin is a roughly spherical molecule with molecular weight of about 42,000 which binds a mole each of nucleotide (ATP or ADP) and divalent cation (Mg^{2+} or Ca^{2+}). Under physiological concentrations of salt and ATP, G-ATP-actin polymerizes into F-ADP-actin with a "beaded" double-stranded, right-handed helical structure, in the groove of which the tropomyosin superhelical structure resides.[25] It is also the F-actin form which combines with myosin to form the contractile actomyosin of living or prerigor muscle tissue and the inextensible actomyosin of muscle in rigor mortis.

The amino acid compositions of actins from various species are shown in Table 6. Roughly 33% of the residues have polar side chains, compared to about 38% for myosin, 50% for tropomyosin B, and 40 to 50% for the troponins.[18] Actin has a relatively high proline content (5.6%), which may account for its relatively low α-helix content (about 30%), compared to the other myofibrillar proteins: about 60% for myosin, 78% for light meromyosin, and 98% for tropomyosin.[18]

Table 7
DISTRIBUTION OF AMINO ACID RESIDUE TYPES IN MYOFIBRILLAR PROTEINS

Protein	Aromatic amino acids[a]	Proline	Basic amino acids	Acidic amino acids	Residues with Polar side Chains
Actin	7.8	5.6	13	14	33
α-actinin	7.7	5.9	12	14	31
β-actinin	7.5	5.7	13	12	39
Tropomyosins B, α and β	3.0	0.1	18	36	50
Troponin C	7.8	0.6	10	33	45[b]
Troponin I	3.7	2.8	20	28	— [b]
Troponin T	5.3	3.5	22	32	— [b]
Myosin	5.3	2.6	17	18	38
H-meromyosin	7.2	3.4	15	15	34
L-meromyosin	2.8	1.0	19	18	39

[a]   Values for aromatic residues do not include tryptophan.
[b]   Average value for troponin (from Bodwell and McClain[18]).

Note: Values represent percentage of total residues. Compiled from data from Lawrie,[6] and Bodwell and McClain.[18]

Actin is evolutionarily highly conserved in amino acid sequence, which probably reflects the strong selective pressure to preserve its three-dimensional structure, thus maintaining essential interaction sites for many associated proteins.[26] In addition to its association with myosin, tropomyosin, and the actinins, actin is associated with a number of minor proteins which cross-link actin filaments to form either bundles or isotropic gels. Thus while the properties of myosins from different muscles have been refined by natural selection to suit the particular requirements of a given muscle, actins appear to be highly conserved. However, the proteins which form gel networks of actin or regulate the dissolution of the gels have evolved specialized properties, so that these actin-binding proteins provide the necessary variations in specialized tissues.

C. Amino Acid Distribution in Myofibrillar Proteins

Table 7 shows the distribution of amino acid residue types in myofibrillar proteins, as percentages of total residues. High proportions (>7%) of aromatic residues are found in actin, the actinins, troponin C, and heavy meromyosin, compared to the other proteins. Proline residues are predominant in actin and the actinins but scarce in tropomyosin B, troponin C, and light meromyosin. Acidic amino acid residues constitute roughly a third of the total residues of tropomyosin and the troponins, and the proportion of polar amino acid residues (acidic plus basic) in tropomyosin and the troponins is roughly 50%. Tropomyosin B has been reported to have the highest "charge density" of any known protein, and contains 80 to 100% α-helix in normal solvents.[18] Although tropomyosin is extracted *in situ* only at high ionic strength, once extracted from the muscle it is soluble in water or solutions of low ionic strength,[6] a property which may be reflecting its high content of polar residues.

Although all the myofibrillar proteins have free sulfhydryl groups, only tropomyosin and the troponins appear to have disulfide groups in their structure,[27] as shown in Table 8. For example, each myosin molecule contains over 40 sulfhydryl residues, but no disulfide bonds, with each myosin head (S-1) containing between 12 to 13 sulfhydryl groups.[28]

Three unusual amino acids have been identified in myosin, namely methylhistidine, ε-N-monomethyllysine and ε-N-trimethyllysine, while N-methylhistidine has been identified in actin; however, their biological significance is not clear.[4]

Table 8
SULFHYDRYL AND DISULFIDE GROUPS IN MYOFIBRILLAR PROTEINS

Protein	Molecular weight	Equivalents per 10^5 gram protein	
		SH	SS
Myosin Rod	220,000	7.3—9	0
HMM	320—360,000	7—10	0
Subfragment 1	—	10	0
Subfragment 2	—	6.4	0
LMM	—	4	0
Actomyosin	—	7.5—8.8	0
Actin	46,000—47,000	7.8—13	0
Actinin	95,000	5.1	
Tropomyosin	70,000	3.1	3.7
Troponin	81,000	5—12.2	4—4.6

Adapted from data cited by Hofmann and Hamm;[27] ranges are given to reflect values from different literature sources.

IV. HYDROPHOBICITY OF MUSCLE PROTEINS

A. Myosin

The question of the effective or surface hydrophobicity of contractile myofibrillar proteins was not given much attention in the past. Early studies on the effects of aging, temperature, and chemical modification of myosin on its ATPase activity, suggested conformational changes in the myosin molecule. However, attempts to demonstrate conformational differences using techniques such as optical rotatory dispersion were unsuccessful,[29] probably due to the large size of myosin in comparison to its active site.

The usefulness of the hydrophobic fluorescence probe 8-anilino-1-naphthalenesulfonic acid (ANS) to study myosin conformation, nucleotide binding, and hydrophobic sites was demonstrated in a series of studies in the 1960s. Duke et al.[30] demonstrated that the fluorescence of a myosin-ANS system developed very quickly, but showed a gradual increase in intensity on prolonged incubation at constant temperature. Similarly, the fluorescence intensity was higher if the myosin had been stored ("aged") prior to incubation with ANS. The increase in fluorescence intensity was attained whether myosin was stored for hours or days at low temperature ("slow" aging) or for minutes at higher temperature ("fast" aging). When the sulfhydryl groups of myosin were modified with p-chloromercuribenzoate (PCMB), the fluorescence intensity of myosin-ANS after incubation for 600 sec, (F_∞) was markedly higher for myosin with increasing extent of modification. However, the fluorescence intensity after prolonged incubation (F_∞) was similar for unmodified and modified myosins, suggesting that the PCMB-induced conformational changes leading to enhanced fluorescence may also be induced by aging. The enhancement in fluorescence by PCMB modification of sulfhydryl groups could be correlated to changes in ATPase activity (Figure 2). With moderate extent of PCMB modification, F_{600} increased gently, while ATPase activity reached a maximal level. However, upon more extensive modification yielding 70 to 90% titration of sulfhydryl groups, F_{600} abruptly increased and plateaued, while ATPase activity disappeared.

Lim and Botts[31] also suggested that conditions favoring moderate interaction of myosin with ANS corresponded to conditions favoring ATPase activity, while extensive ANS binding resulted in loss of enzymatic activity. They demonstrated that slow aging at 4° or brief exposure to increased temperatures from 25 to 50° resulted in progressively declining ATPase activity and progressively increasing bound ANS fluorescence. These conditions were sug-

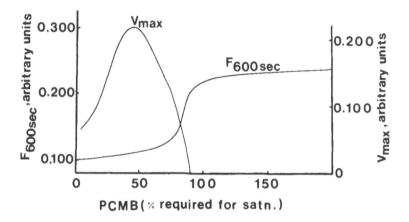

FIGURE 2. Changes in fluorescence intensity at 600 sec on adding ANS to myosin (F_{600}) and ATPase activity (V_{max}), as a function of varying extents of the sulfhydryl modification by p-chloromercuribenzoate (PCMB). (Adapted from Duke, J. A., McKay, R., and Botts, J., *Biochim. Biophys. Acta*, 126, 600, 1966. With permission.)

gested to induce progressive structural changes in myosin, increasing the number of hydrophobic regions able to bind ANS. The nonpolar or hydrophobic nature of ANS binding was implied since the wavelength of maximum emission (475 nm) was similar to that for ANS in a nonpolar medium such as octanol.

From a linear plot of $1/F$ vs. $1/D$, where F was the observed fluorescence of the ANS-myosin complex at D concentration of ANS, Cheung and Morales[32] estimated that there was one ANS-binding site per myosin molecule, and they calculated an average intrinsic dissociation constant, K, for the ANS-myosin complex. At 20°, K was $4.93 \times 10^{-5} M$, while at 8.7°, it was $5.76 \times 10^{-5} M$. From the temperature dependence of the dissociation constant, they roughly estimated values of ΔH and ΔS to be 2.28 kcal/mol and 26 e.u. at 20°C, respectively, consistent with hydrophobic forces being significant in the ANS-myosin interaction. Their data indicated quantum yield and wavelength of maximum emission of 0.48 and 468 nm for myosin-bound ANS, compared to 0.004 and 514 nm for the corresponding values of ANS in water, and 0.63 and 464 nm for ANS in n-octanol. These results strongly support the hypothesis of hydrophobic interactions in the binding of ANS to myosin. The binding site was postulated to be at the myosin head, since ANS did not appear to bind to light meromyosin, while it did bind to heavy meromyosin with fluorescence enhancement of between 30 to 50% that of myosin-ANS fluorescence.

The application of ANS to study conformational changes in myosin in regions proximal to the active centers was investigated by Cheung and Morales,[32] who demonstrated the quenching of intrinsic fluorescence of myosin by its interaction with ANS. In the absence of ANS, excitation of myosin at 295 nm yielded a fluorescence emission spectrum with a peak at 345 nm, arising from tryptophan residues. Upon addition of increasing ANS to myosin, excitation at 295 nm resulted in progressive quenching of the tryptophan fluorescence at 345 nm and appearance of a new peak at 475 nm, resulting from transfer of excitation energy from the myosin chromophores to the bound ANS. Efficiency of energy transfer decreased slightly upon moderate extent of modification of myosin with PCMB, and the emission wavelength maximum underwent a slight blue shift (to 465 nm), suggestive of a change in the localized environment of the bound ANS to a more nonpolar type. Upon more extensive PCMB modification, a large decrease in energy transfer efficiency and greater blue shift (to 463 nm) were observed, accompanied by inactivation of ATPase.

According to Förster's theory,[33] the transfer efficiency is very sensitive even to small localized conformational changes, since it is proportional to an inverse sixth power of the distance between the acceptor and donor in the transfer, and also depends on the mutual orientation of the two chromophores. Since the ANS-binding site was suggested to be on the myosin head, and the transfer would be possible only at a separation distance of roughly 50 Å or less, the tryptophan residue(s) responsible for the transfer were also suggested to be located on or near the myosin head.[32] Perturbations affecting the activation or deactivation of ATPase may have caused small localized conformational changes which were reflected in reduced transfer efficiency from myosin chromophores to bound ANS.

Evidence for a reversible conformational change in a myosin-localized structure during its enzymatic catalysis of ATP hydrolysis was demonstrated from the decrease in transfer efficiency from myosin chromophore to bound ANS in the presence of ATP.[34] Thus, fluorescence of myosin-bound ANS excited at 280 nm showed a small reversible decrease in the presence of ATP. The decrease in energy transfer was correlated with the kinetics of ATP hydrolysis to ADP, and restoration of fluorescence from energy transfer occurred upon depletion of ATP. Loss of ATPase activity by urea, alkalines pH, N-ethylmaleimide or iodoacetamide treatment were also associated with myosin conformational changes which were reflected in the altered fluorescence properties of the myosin-ANS complex. In analogy to the earlier results of PCMB modification, low concentrations of urea (<2 M) or slightly alkaline pH (8 to 10) resulted in slight increases in myosin-bound ANS fluorescence and maximal activation of the ATPase enzyme activity. Increasing urea concentration to 3 M and pH to 10 or above, resulted in a steep increase in myosin-bound ANS fluorescence and rapid inactivation of ATPase; under these conditions, the wavelength of maximum emission was 462.5 to 463 nm, compared to 468 nm for the native myosin-ANS complex, suggesting a large change in the polarity of the environment of bound ANS. At even higher urea concentration (6 M) or pH (11.5), the fluorescence was drastically reduced, even to levels below that of control myosin, suggesting possible destruction of the hydrophobic site under these drastic conditions. The extent of denaturation by high urea could be reduced by the addition of ATP prior to incubation with urea, suggesting the stabilizing effect of ATP on myosin structure. These results indicated the usefulness of the ANS probe to detect both the small and localized conformational changes associated with activation of ATPase and during subsequent hydrolysis of ATP, as well as the larger, more extensive conformational changes associated with ATPase inactivation.

As an alternative to ANS as the fluorescent hydrophobic probe, Borejdo[35] used *cis*-parinaric acid (CPA) to further investigate the hydrophobic sites on the surface of myosin and its fragments. Contrary to the findings using ANS, binding of CPA had no effect on the ATPase activity of myosin. At both high and low ionic strength, myosin possessed 1.3 high affinity binding sites for reversible binding of CPA, with a dissociation constant K_D of about 10^{-7} M. Neither the binding of actin nor the binding of nucleotide, nor the use of myofibrils in place of myosin had any significant effect on the stoichiometry or affinity of CPA binding, although the rotational relaxation time of bound CPA was increased in the presence of actin and in the myofibrils. F-actin itself only bound CPA weakly ($K_D > 10^{-5}$ M), compared to myosin binding of CPA($K_D = 10^{-7}$ M). The data indicated that CPA was bound to a hydrophobic pocket at the myosin head, between the S-1 heavy chain and alkali light chain 1, and that this site was insensitive to changes occurring at the actin-myosin interface or to nucleotide binding.

Hydrophobic surface properties of myosin were studied by Pinaev et al.[36] using affinity partitioning in dextran-polyethylene glycol-water biphasic systems. In 0.5 M KCl at pH 7.5, rabbit muscle myosin was increasingly transferred to the polyethylene glycol (PEG)-rich upper phase when long chain fatty acyl esters of PEG were present in that phase, yielding increases in the partition coefficient of myosin from one hundred- to one thousand-fold. This indicates the considerable hydrophobic character of the myosin surface. The partition

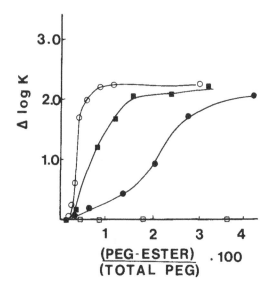

FIGURE 3.   Effect of polyethylene glycol fatty acid esters on partitioning of myosin in biphasic systems of dextran (4.5%), polyethylene glycol (4.5%), and aqueous buffer (0.5 M KCl, 5 mM sodium phosphate buffer, pH 7.5). The fatty acid esters are caprinate (□), laurate (●), myristate (■), and palmitate (○). (Adapted from Pinaev, G., Tartakovsky, A., Shanbhag, V. P., et al., *Molecular & Cellular Biochemistry*, 48, 65, 1982. With permission.)

coefficient increased in a sigmoidal fashion with increasing proportion of PEG-ester to total PEG, suggesting a cooperative type of binding of hydrophobic groups by myosin, which is in agreement with the cooperative nature of the ionic strength-dependent self-association of myosin molecules. The effect of fatty acyl ester on partition increased in the order of increasing chain length of the fatty acyl group: laurate (C_{12}) < myristate (C_{14}) < palmitate (C_{16}) (Figure 3). No affinity of myosin for the caprinate (C_{10}) ester was demonstrated, in contrast to the behavior of other proteins such as human serum albumin and β-lactoglobulin.

Partition studies on the effects of ionic strength, pH, and temperature, indicated that the conditions favoring tendency for myosin molecules to self-associate and form filaments, were also those yielding increase in the number or hydrophobicity of sites on myosin, responsible for the preferential partitioning into the phase with PEG-palmitate. Thus, the affinity of myosin for PEG-palmitate increased with (1) decreasing ionic strength (from 0.6 M to 0.2 M KCl at pH 7.5, at 4 or 20°C), (2) higher temperature (20° vs. 4°C), and (3) decreasing pH (from 8.5 to 5.6 in the presence of 0.2 M KCl or 0.5 M KCl). Pinaev et al.[36] concluded that hydrophobic binding makes a substantial contribution to the stability of myosin filaments at least in the range of salt concentration of 0.2 to 0.5 M KCl.

B. Actin

The importance of hydrophobic interactions in the elongation stage of actin polymerization from globular G-actin to fibrous F-actin has been suggested by the entropy-driven nature of this process.[37] Asakura et al.,[38] in a study on the effect of temperature on the equilibrium state of rabbit muscle actin solutions, concluded that in the 0 to 20°C temperature range, the G-F transformation is favored by higher temperatures, while the F-G transformation is favored by lower temperatures. These authors calculated a large positive enthalpy change ΔH of the order of 10 kcal per mol of actin for the G-F transformation. Since the G-F transformation occurs spontaneously at the higher temperature, the free energy change ΔG must be negative, and the change in entropy ΔS must be a large positive value, to meet the criterion of $\Delta G = \Delta H - T\Delta S > 0$. They postulated that the G-F transformation was a transition of hydrated G-actin into dehydrated F-actin plus free water, with net increases in both entropy and enthalpy. These results were also supported by Gordon et al.[39] who showed even more highly positive ΔH and ΔS of polymerization of actin from *Acanthamoeba*, and attributed the phenomenon to a conformational change in the actin monomer during polymerization, with expulsion of bound water.

Two different hydrophobic fluorescence probes were used to study the hydrophobic-binding sites of G-actin, F-actin, and the nonpolymerizable complex formed between G-actin and deoxyribonuclease (DNase) I.[37] The probe 2-(N-methylanilino)naphthalene-6-sulfonic acid (MANS) bound to G-actin at a single high affinity hydrophobic site, with a dissociation constant of 41 μM. G-actin bound MANS excited at 365 nm contained two emission peaks of enhanced intensity, one at 490 nm and the other at 430 nm, compared to a single smaller emission peak at 507 nm for free MANS in buffer. The double-peaked emission of MANS-G actin solutions was also produced for free MANS in organic solvents (glycerol, acetone, propanol, ethanol, or ethylene glycol as 95% (v/v) solutions with water). In these solvents, increasing fluorescence intensity especially at the lower wavelength peak (430 nm) was observed with increasing nonpolar nature of the solvent.

Salt-induced polymerization of MANS-G actin by KCl or $MgCl_2$ produced increased emission intensity at all wavelengths, but especially at 430 nm, suggesting a more hydrophobic or nonpolar environment of the bound MANS after polymerization. It was postulated that the MANS-binding site may be a hydrophobic site of contact of one actin protomer with another in the F-actin filament.

The second type of probe used by Burtnick and Chan[37] was 9-anthroyl choline bromide (9AC). It also bound to G-actin at a single hydrophobic site, yielding enhanced fluorescence at emission wavelength maximum of 495 compared to 500 nm for free 9AC in buffer. Unlike MANS-G actin, however, salt-induced polymerization of 9AC-G actin by KCl decreased the fluorescence intensity, while polymerization by $MgCl_2$ slightly increased the fluorescence intensity. The salt-dependent changes upon polymerization, support earlier work that KCl and $MgCl_2$ induced polymers may not be identical in nature. It was also postulated that although MANS and 9AC were each bound to one hydrophobic site on actin, those two sites were probably not identical. Though both are fluorescent hydrophobic probes, 9AC is cationic while MANS is anionic, and binding to actin may have been modulated by these charges.

C. Actomyosin, Meat Sols, and Salt Extracts

The role of hydrophobic bonding in gelation of fish flesh paste was investigated by observing the fluorescence of ANS added to actomyosin extracts from flatfish.[40] Upon excitation at 365 nm, enhanced fluorescence emission at 470 nm was observed when actomyosin had been heated at 40, 60, and 90°C prior to addition of ANS, compared to unheated actomyosin. The highest fluorescence intensity was observed for 60°C-heated actomyosin, while the enhancement in 40°C-heated samples was depressed in the presence of saccharides, glycerol, and ethylene glycol.

"Setting" or "suwari" is a phenomenon observed in fish sol (fish mince with added sodium chloride), either upon standing at low temperature or upon heating at 40°C. An elastic gel is formed by setting. However, it is observed only in some species of fish (e.g., flatfish, sardine, jack mackerel, and yellowtail), but not in other species of fish (e.g., carp and dolphinfish), and not in species such as pork, beef, and chicken.[41,42] Fluorescence intensity of actomyosin-ANS complexes or myosin-ANS complexes of the different species indicated that fluorescence intensity was invariably higher for the easily-setting species and that the intensity was also increased more intensively after heating of those species.[42,43] Suppression of further increases in intensity upon prolonged heating was postulated to be due to scattering of the excitation light by protein aggregates. It was suggested from these results that before heating, surface hydrophobic interactions presumably related to intra-molecular interactions are more abundant in the easily setting species than in the species which do not set; upon heating, the hydrophobic interactions increase more intensively, presumably related to the formation of intermolecular interactions.

Niwa et al.[44] proposed that hydrophobic bonding was important for regulating the setting of fish flesh sol to gel, since both hydrophobic bonding and setting were enhanced by heating, presence of salt, or low concentration of urea. On the other hand, upon cooling, hydrogen bonding was suggested to be a contributor to final elasticity of the gel.

The contribution of aromatic amino acid residues to the setting phenomenon was investigated by arylation of myosins and actomyosins from different species.[45] The importance of aromatic residues was suggested by the correlation of a higher extinction coefficient at 280 nm for the myosin heavy chains from easily-setting species such as flatfish, jack mackerel, and yellowtail ($E_{1\,cm}^{1\%}$ of 21, 31, and 25, respectively), compared to the lower values for myosin heavy chains from carp, pig, and hen ($E_{1\,cm}^{1\%}$ of 8, 8, and 9, respectively). However, the harsh alkaline conditions used for isolation of the myosin heavy chain fraction was speculated to influence the exposure of amino acid residues of the molecule. Thus, these researchers artificially introduced aromatic side chains onto chicken and pig muscle myosins and actomyosins, to confirm the importance of these residues in the setting phenomenon.

Arylsulfonates (potassium β-naphthoquinone-4-sulfonate and sodium β-naphthoquinone-4,7-disulfonate) which react with NH$_2$-residues, and arylsulfonyl chlorides (1-dimethyla-mino-5-naphthalenesulfonyl chloride, p-toluenesulfonyl chloride and α-naphthalenesulfonyl chloride) which react with OH-, SH-, and NH$_2$- residues, were used to modify the proteins. The results showed that arylated hen and pig myosins and actomyosins had increased viscosity and gelled readily at 40°C, while the unmodified controls did not. The setting or gelation of aryl-modified proteins was induced even at 4°C but was suppressed by the addition of sucrose. Thus, the muscle proteins from hen and pig were induced to behave like those from easily-setting species, by the introduction of hydrophobic aromatic functional groups. It was interesting to note that some of the arylated actomyosins became insoluble yet maintained biological activity and also gelling ability; it was suggested that the insolubili-zation was induced not by unfolding or denaturation, but by increase in the hydrophobic nature of the molecular surface through the introduction of the aryl groups.

In a similar study, dolphinfish flesh sol, consisting of the salted washed mince, was induced to behave like the sols from easily-setting species simply by addition of arylating agents to the flesh sol,[46] demonstrating the importance of the aromatic groups to setting of intact flesh sol, as well as the extracted protein fractions.

To demonstrate the importance of nonpolar or hydrophobic aliphatic amino acid residues in setting, slow-setting species (chicken and beef) were modified by ethylsulfonation of actomyosin sols or addition of ethylsulfonyl chloride to flesh sols.[47] Ethylsulfonated chicken actomyosin lost its salt-solubility as well as ATP-sensitivity, and readily gelled at 40°C; the gelation was suppressed in the presence of sucrose. Similarly, the breaking stress increased and expressible water decreased for 40°C-heated flesh sols containing 0.3 or 0.1% ethyl sulfonyl chloride, compared to the control 40°C-heated flesh sols in the absence of the reagent. Preheating at 40°C of the flesh sols containing ethylsulfonyl chloride, prior to further heating at 90°C, yielded an increased breaking stress compared to sols heated only at 90°C, whereas no effect of preheating was observed for control sols. These results suggested that both aliphatic and aromatic amino acid residues were related to the setting phenomenon, and that hydrophobic interactions are crucial in this phenomenon.

The hydrophobic nature of proteins in salt extracts of beef, fish, and chicken was investigated using an aliphatic fluorescent hydrophobic probe (CPA) as well as an aromatic fluorescent probe (ANS).[48-50] The salt extracts of these meat samples probably contained both water- and salt-soluble proteins, mimicking the conditions normally existing in comminution of meat with salt and ice. Hydrophobicity of the proteins in these crude extracts was determined as the initial slope (S_O) from plots of fluorescence intensity vs. protein concentration, according to the method of Kato and Nakai.[51]

Figure 4 shows typical data of the hydrophobicity of proteins in salt extracts from beef muscle, determined using the CPA probe (S_OCPA), plotted as a function of heating temperature.[48] The simultaneous changes in protein solubility upon heating are also shown, determined as the protein remaining in the supernatant after high speed centrifugation (27,000 × g, 30 min). Heating at temperatures from 50 to 70°C progressively resulted in increased S_OCPA and decreased solubility. At 70°C, the proteins in the salt extracts had two- to three-fold increases in their surface hydrophobicity as reflected by S_OCPA values, and decreases in solubility to 20 to 25% of the values for the unheated control extracts. Further heating to temperatures higher than 70°C resulted in readily visible coagulation of proteins in the extracts, accompanied by slight decreases in both surface hydrophobicity and solubility. Extracts which were heated to 75°C in the presence of 0.05% sodium dodecylsulfate (SDS) had effective hydrophobicity values similar to those extracts heated in the absence of SDS; however, SDS-treated extracts did not coagulate after heating. Similar results were obtained for S_OANS, the hydrophobicity determined using ANS as the fluorescent probe.[49] These results suggest that heating resulted in increasing exposure of both aromatic and aliphatic residues of the muscle proteins, reflected in the increases in both S_OANS and S_OCPA.

In general, salt-extractable proteins from beef (top round), chicken (breast muscle), fish (rockfish or ling cod fillets), and pork (lean ham trim) all showed qualitatively similar effects of increased surface hydrophobicity upon heating. However, quantitative differences between species, between fresh and frozen samples, and between hand- and mechanically-deboned samples, were apparent both before and after heating of the extracts (Table 9).[52] Among hand-deboned samples, S_OANS was highest for salt-extractable proteins from fish muscle and lowest from chicken, while S_OCPA was highest from fish and pork samples and lowest from chicken. Salt-extractable proteins from samples after frozen storage at −10 or −20°C for up to 32 weeks had generally higher S_OANS and S_OCPA. Extracted proteins from mechanically-deboned samples, and from kidney and tripe samples, also had generally higher S_OANS and S_OCPA than their hand-deboned skeletal muscle counterparts.

Heating of salt-extractable proteins increased S_OANS and S_OCPA in all samples; however, the extent of increase differed, as reflected by hydrophobicity of heated extracts as well as the ratio of heated/unheated hydrophobicity values. Hand-deboned chicken samples showed the greatest increase in S_O after heating (10-fold and 5-fold increases in S_OANS and S_OCPA heated/unheated ratios, respectively), whereas mechanically deboned samples, frozen samples, and kidney and tripe samples all showed much smaller changes by heating, with 1.2 to 2-fold ratios of heated/unheated hydrophobicity values. For the mechanically deboned, kidney, and tripe samples, differences in surface hydrophobicity may arise from their different composition of proteins in the salt extracts, compared to extracts from hand-deboned skeletal muscle. For the frozen samples, structural changes in the proteins are likely to be the basis for the higher surface hydrophobicity of salt-extractable proteins.

In addition to surface hydrophobicity of salt-extractable proteins, Table 9 also shows the ranges of other physicochemical properties, including salt-extractable protein content, sulfhydryl content and solubility of the proteins, and pH, % moisture, fat, and protein in the mince. Differences in these properties as well as in functional properties (Table 10) are apparent.[52] The relationships between these functional and physicochemical properties were further explored by the approach of quantitative structure activity relationship using multivariate data analyses,[53] as described later in this text (Section VI).

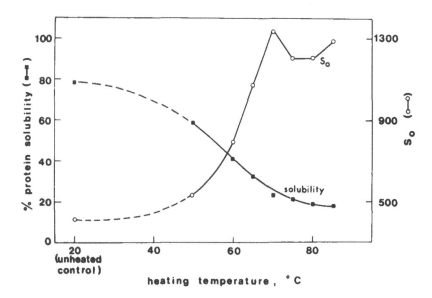

FIGURE 4.   Changes in hydrophobicity (S_o) and % solubility of proteins in salt extracts from top round beef, as a function of heating temperature. (From Li-Chan, E., Nakai, S., and Wood, D. F., *J. Food Sci.*, 49, 345, 1984. With permission.)

The effects of acidic vs. neutral pH on the surface hydrophobicity of beef and rockfish salt-extractable proteins were investigated using both ANS and CPA hydrophobic probes.[49] Generally, salt extracts of beef and rockfish proteins adjusted to pH 5.5, had higher S_oANS than extracts at pH 7.0, suggesting greater unfolding of the protein molecules to expose aromatic groups as the pH approached their isoelectric point. The quantum yield of the ANS probe has been documented to be insensitive to pH change in the pH 2 to 8 range,[54] suggesting that the effect of pH on S_oANS of the muscle proteins was due mainly to protein structural changes or alteration in binding sites or hydrophobicity of the environment of the bound ANS probe at the different pH conditions. On the other hand, the effects of pH on S_oCPA of salt-extractable proteins were less clear. While S_oCPA of beef proteins paralleled S_oANS in being generally higher at pH 5.5 than at pH 7.0, for rockfish extracts, S_oCPA was lower at pH 5.5 than at neutral pH. These results implied that for rockfish, although aromatic hydrophobic groups (determined as S_oANS) were more accessible at pH 5.5 than at pH 7.0, the converse was true for aliphatic hydrophobic groups (determined as S_oCPA), which became less accessible with decreasing pH. Differences in the exposure of aromatic and aliphatic amino acid residues have been reported;[55] the bulkier nature of the aromatic groups may induce their preferential orientation towards the surface of the molecule while the smaller aliphatic side chains may become buried within the hydrophobic core.

The changes in surface hydrophobicity of salt-extractable proteins as a function of different postmortem storage times at above-freezing temperature were monitored for beef neck muscle[56] and chicken breast muscle.[50] For beef neck muscle sampled at 0, 1, 3, and 10 days post-slaughter, the concentration of salt-extractable proteins decreased progressively with increasing time up to 3 days, then increased slightly at 10 days (Figure 5). Surface hydrophobicity determined as S_oCPA increased progressively with postslaughter time, being highest at 10 days (Figure 6). Similar results were obtained for chicken breast muscle analyzed at 0, 2, and 7 days postslaughter time. Both salt-extractable protein concentration and surface hydrophobicity determined as S_oANS and S_oCPA were higher at 7 days than at 0 or 2 days. It was suggested that the natural aging process in meats results in increased accessibility or exposure of both aliphatic and aromatic hydrophobic sites in the salt-extracted proteins.

Table 9

SUMMARY OF RANGE OF PHYSICOCHEMICAL PROPERTIES FROM VARIOUS SPECIES, INCLUDING FRESH AND FROZEN, HAND-DEBONED, AND MECHANICALLY DEBONED SAMPLES[52]

Properties	Beef	Cod (Ling)	Pork fresh	Pork frozen	Chicken fresh	Chicken frozen	MD-pork fresh	MD-pork frozen	MD-chicken fresh	MD-chicken frozen	Beef kidney	Beef tripe
Salt Extracts[a] Protein extractability (g/10 g sample)	0.8—1.2	1.1—1.2	1.3—1.5	0.9—1.0	1.2—1.5	1.4—1.5	0.8	0.7	0.5—0.6	0.4—0.6	0.9	0.1
S_oANS unheated	100	200	100—130	120—160	65	100	120	120—160	100	110—125	ND	ND
heated[b]	300	300—400	400—500	450—500	600—700	500	200	200	400	300	ND	ND
(heated/unheated) ratio	3	1.5—2	4	4	10	5	2	1.5	4	2.5—3	ND	ND
S_oCPA unheated	350—500	500—650	550—700	550	280—350	450—500	600—700	750—800	900—1000	800—1000	800—1200	800
heated[b]	1200	900—1000	1500	1200	1500	1000	1000	850—1000	1400	1200—1400	1600—1900	920
(heated/unheated) ratio	3	1.5—2	2.5—3	2	5	2	1.5	1.2	1.5	1.5	2	1.2
SH unheated (μM/g)	60—80	70—90	80—90	80	90—100	80	65	63	70—75	70—75	ND	ND
heated[b] (μM/g)	50—60	60—80	80—90	70	80	70	55	54	70—75	70—75	ND	ND
% solubility unheated	80	85	90	83	85—90	82	90	80	90	80	85	85
heated[b]	30—50	30—50	12	7	5	6	18	13	10	10	30	85
Mince pH of suspension	5.7—6.0	6.6	5.7	5.8	5.7—6.0	5.8	6.6	6.8	6.6	6.7	ND	ND
% moisture in mince	74	86	75	75	76	76	61	62	60—65	60—65	79	85
% protein in mince	18—20	7—10	13	13	19—22	19—22	14	14	11—13	11—13	11—12	ND
% fat in mince	1—3	<0.5	2.5	2.5	0.5	0.5	25	25	16—20	16—20	0.6	ND

Note: ND, not determined. MD, mechanically deboned.

[a] Extracting buffer containing 0.6 *M* NaCl, 1 m*M* MgCl₂, 0.01 *M* Na phosphate, pH 6.5 to 7.0.

[b] Heated at 40/80°C or 80°C.

Table 10
SUMMARY OF RANGE OF FUNCTIONAL PROPERTIES FROM VARIOUS SPECIES, INCLUDING FRESH AND FROZEN, HAND-DEBONED, AND MECHANICALLY DEBONED SAMPLES[52]

	Beef	Cod (Ling)	Pork		Chicken		MD-pork		MD-chicken		Beef kidney	Beef tripe
			fresh	frozen	fresh	frozen	fresh	frozen	fresh	frozen		
Gelation[a]												
mince gel strength, N	9—17	5—6	11	5—15	12—15	13	11	3—4	5	3—5	ND	ND
mince cook loss, %	5—8	0	6	4	5	2—4	2	3—4	10—30	10—35	ND	ND
extract gel strength, N	0.025	0.01	0.015—0.02	0.01—0.02	0.03—0.05	0.03—0.05	0.010—0.015	0.010—0.015	0.03	0.03	ND	ND
Water Binding, g H_2O/g sample												
AM_w	0.1—0.3	1.0—1.1	0.2—0.3	0.2—0.3	0.2—0.3	0.1—0.2	0.3—0.4	0.3—0.35	<0.1—0.2	<0.1—0.2	0.1—0.3	0.8
RM_w	0.2—0.3	0.5—0.6	0.3—0.4	0.3—0.4	0.4—0.6	0.3—0.4	0.3—0.4	0.3—0.4	<0.1—0.2	<0.1—0.2	<0.1	0.7
AM_b	0.2—0.4	0.3—0.4	0.3—0.4	0.3—0.4	0.2—0.4	0.2—0.4	0.3—0.4	0.3—0.35	<0.1—0.2	<0.1—0.2	0.1—0.2	0.4
RM_b	0.3—0.5	0.15—0.20	0.4—0.6	0.4—0.6	0.5—0.8	0.4—0.5	0.4—0.5	0.4—0.5	0.2—0.3	0.15	<0.1—0.1	0.4
Emulsifying Capacity, g oil/30 mg protein												
unheated extract	45—65	50—70	50	50	50	50	56	52	60	45—50	50—80	ND
heated extract[a]	15	30—50	25	25	25	25	20	20—30	30	25—30	ND	ND
Fat Binding Capacity, mℓ oil/g protein												
unheated extract	3—4	4—10	5—7	3—5	13—16	7—10	6	3—5	6	5—7	ND	ND
heated extract[a]	2	3—14	4—6	3—5	3—5	2	2—3	2	2	3	ND	ND

Note: ND, not determined. MD, mechanically deboned.

[a] Heated ≡ 40/80°C or 80°C.

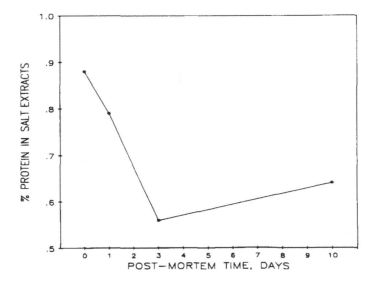

FIGURE 5. Effects of postmortem time on % protein in salt extracts from beef neck muscle. (Adapted from Wood, D. F., Campbell, C. A., Li-Chan, E., and Nakai, S., *Abstr. 38, Can. Inst. Food Sci. Technol. 27th Ann. Conf.*, Vancouver, Canada, 1984.)

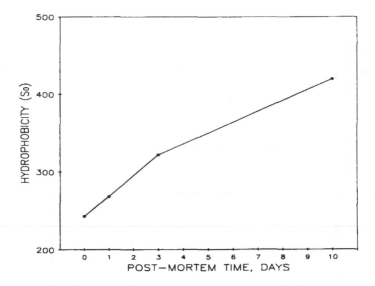

FIGURE 6. Effects of postmortem time on hydrophobicity (S_o) of proteins in salt extracts from beef neck muscle. (Adapted from Wood, D. F., Campbell, C. A., Li-Chan, E., and Nakai, S., *Abstr. 38, Can. Inst. Food Sci. Technol. 27th Ann. Conf.*, Vancouver, Canada, 1984.)

D. Intact Muscle

Intrinsic protein fluorescence of strips of meat directly placed on the front surface of a quartz cell was used to study the conformational changes in muscle proteins of chilled cured beef during heating, as reflected by the aromatic hydrophobic amino acid residues; these fluorescence studies were related to data obtained by differential scanning calorimetry.[57]

The fluorescence spectra of the samples at each temperature were characterized by three

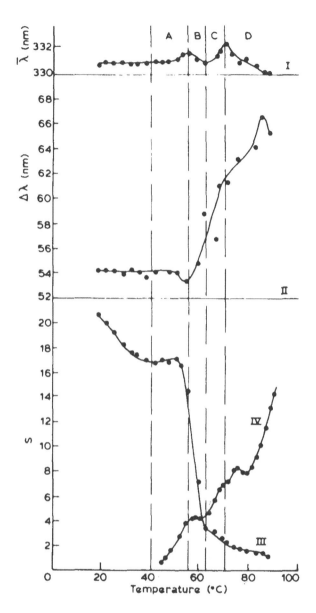

FIGURE 7. Fluorescence parameters of cured meat as a function of heating temperature. I: spectrum position ($\bar{\lambda}$), II: spectrum width ($\Delta\lambda$), III: area under the spectrum (S), and IV: microcalorimetric thermogram. (From Oreshkin, E. F., Borisova, M. A., Tchubarova, G. S., et al., *Meat Sci.*, 16, 297, 1986. With permission.)

parameters; spectrum position ($\bar{\lambda}$), spectrum width ($\Delta\lambda$) and area under the spectrum (S) which is proportional to the fluorescence quantum yield (Figure 7). In addition, contribution of the discrete classes of tryptophan residues to total emission was estimated (Figure 8). The four classes were designated S, I, II, and III. Spectral classes S and I corresponded to interiorly located, water-inaccessible residues of differing mobility, and classes II and III corresponded to surface residues which contact bound (II) and free (III) water molecules. Figures 7 and 8 show that four temperature regions A to D could be distinguished, characterized by distinct changes in the proteins.

In region A (40 to 55°C), there was a slight shift of $\bar{\lambda}$ to longer wavelengths (Figure 7-

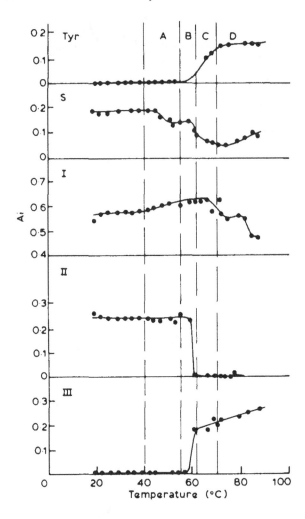

FIGURE 8.   Contribution of emission of tyrosine (Tyr) and tryptophan (S, I, II, and III) residues to total emission spectrum of cured meat, as a function of heating temperature. (From Oreshkin, E. F., Borisova, M. A., Tchubarova, G. S., et al., *Meat Sci.*, 16, 297, 1986. With permission.)

I), and a slight increase in S (Figure 7-III), which was accompanied by a peak in the thermogram (Figure 7-IV). These changes were attributed to a decrease in a less mobile type S of tryptophan residues (Figure 8-S) and an increase in the more mobile type I residues (Figure 8-I), signifying a loosening of the meat protein structure.

Region B (55 to 61°C) was characterized by a slight decrease in $\bar{\lambda}$ to shorter wavelengths (Figure 7-I) and a pronounced decrease in spectrum area S (Figure 7-III), possibly reflecting major structural coagulation-type changes in the meat proteins.

In region C (62 to 70°C), $\bar{\lambda}$ was again shifted to longer wavelengths (Figure 7-I), accompanied by another peak of heat sorption (Figure 7-IV), major declines in types S, I, and II spectral classes contribution, and increases in type IV and tyrosine contributions to total emission (Figure 8). This region is indicative of major denaturation, loosening, and hydration of the protein structure, and exposure of previously buried tryptophan residues coming into surface contact with free-relaxing water molecules.

The last temperature region D (>70°C) was similar to region B in that a coagulation process had the effect of transferring some tryptophan residues back to the inaccessible regions (increase in type S, Figure 8-S), although further denaturation was still occurring, as reflected by the continued increase in type III class (Figure 8-III). The net effect was a continued decrease in the area under the spectrum S (Figure 7-III), shift in $\bar{\lambda}$ to shorter wavelengths (Figure 7-I) and another peak in the thermogram (Figure 7-IV). These results demonstrate the changes in the accessibility of hydrophobic aromatic residues upon heating, which could be related to coagulation and changes in the thermogram of the muscle strips.

V. FUNCTIONAL PROPERTIES AND MUSCLE PROTEIN STRUCTURE — QUALITATIVE ASPECTS

The characteristic quality of comminuted processed meat products including emulsion, particulate, sectioned, formed, and restructured types of products, is dependent on the functional properties of the protein matrix.[58] During the comminution process, the meat is reduced to a fine particulate state in the presence of salt and ice which induce swelling, water binding, and partial extraction of salt-soluble protein components.[59] Subsequently, the fat component is dispersed or "emulsified" within the protein sol matrix, followed by heat processing to set or "gel" the emulsion and protein matrix. Thus the properties of the functional protein matrix in comminuted systems have been described in terms of emulsification and/or binding and gelation performance,[58] with the final objective of yielding a product with the desired textural qualities, maximum cook yield and minimum fat coalescence or "fat cap" production.

It should be pointed out that a meat "emulsion" is not a true emulsion, which should be strictly defined as a biphasic system in which an immiscible liquid is dispersed as small droplets (dispersed phase) in another immiscible liquid (continuous phase).[60] In fact, a meat emulsion is not biphasic, but rather a multiphase system consisting of a complex colloidal aqueous system or matrix of salts, proteins, and other soluble components, in which are dispersed solid components including insoluble proteins and fat particles. The stability of meat emulsions, especially after heat processing, depends not only on the minimization of flocculation or coalescence of the dispersed phase, but also on the formation in the continuous phase of a stable protein matrix gel which entraps water as well as fat. It has been suggested that in the case of the fat dispersion in a gel matrix (gel-type emulsion), it is the formation of a stable protein matrix, rather than the formation of an interfacial film around the dispersed phase, which keeps fat particles from coalescing by restricting mobilization of the fat.[10] It has been suggested also that the comminuted meat system would be more appropriately described as "minced" meat or meat "batter" system, rather than emulsion. Yet, as Acton et al.[59] pointed out, examination of the product during processing shows that there are some characteristics in the system which permit retaining the optional term "emulsion". Thus, the fat particles appear to be dispersed and are subdivided during the comminution process into smaller particles, with a proteinaceous membrane deposit around the fat particle or globule interface. In addition, fat liquefaction has been noted to occur at temperatures lower than those required for optimum gel matrix formation. Thus the importance of hydrophobic properties which determine ability to form an interfacial protein film may still be important as a stabilizing factor in comminuted meat systems.[61]

With the above points in mind, the following discussion presents some qualitative aspects of the relationships between muscle protein structural properties and the functional properties which have been suggested to play important roles in the quality of comminuted or restructured meat products, namely emulsifying, water-binding, fat-binding, and gelling properties.

A. Emulsification
Following the description of a wiener batter as an emulsion by Hansen,[62] many workers

have used oil titration methods measuring emulsifying capacity to evaluate the functionality of meat as well as nonmeat ingredients for use in comminuted meat products. The model system described by Swift et al.[63] has probably been the most popular method for measuring the emulsifying capacity, and basically determines the volume of oil that can be emulsified by the protein in solution, before phase inversion or collapse of the emulsion occurs. Emulsifying activity[64] and emulsifying activity index[65] based on height of the emulsion layer and turbidimetric properties of the emulsion, respectively, have also been suggested as alternatives for assessing emulsifying properties. These methods evaluate emulsifying activity based on the properties of emulsions formed by mixing or blending predetermined amounts of oil and protein solution. An additional parameter frequently used to evaluate emulsifying properties is the stability of the emulsion, typically after subjecting to standing for prolonged periods of time, low speed centrifugation and/or heating.

Commonly, the conditions for determining the emulsifying properties of proteins are chosen arbitrarily and may differ widely between various investigators. The nonstandardized conditions used for the determination of this functional property make valid comparison of data from various studies difficult, if not impossible. Examples of conditions which were found to affect the data include type of oil, oil phase volume, equipment design, shape of the emulsifying container, speed of blending, rate of oil addition, temperature, pH, and presence of nonprotein components including salts and sugars.[9,66,67] These sources of variation are superimposed on the effects of protein concentration and composition. Nevertheless, the model system studies for fat emulsification using purified proteins, heterogenous protein extracts, as well as meat homogenates, have yielded some fundamental information on the interactions of proteins and lipids in emulsion systems.

Evaluation of the functionality of individual muscle proteins in meat emulsions has suggested that the salt-soluble proteins are the dominating fraction which contributes to good emulsifying capacity. Hansen[62] noted from microscopic observations that myosin and actomyosin appeared to concentrate at the fat globule surfaces to form a stabilizing membrane. Swift et al.[63] reported that salt-soluble proteins were more effective for emulsion preparation than proteins extracted by water. The meat proteins had improved emulsifying capacity in the presence of increasing salt concentrations, especially at pH approaching or lower than 5.4, suggesting a salt-induced shift in the isoelectric point which resulted in enhanced emulsifying capacity by maintaining more protein in solution at the low pH. Baliga and Madaiah[68] emphasized the importance of myosin in stabilizing meat emulsions prepared from mutton. Hegarty et al.[69] ranked the muscle protein fractions from beef muscle in the following order of decreasing emulsifying capacity in the presence of 0.3 M salt: myosin, actomyosin, sarcoplasmic proteins, and actin; in the absence of salt, actin had superior emulsifying properties but the significance of this is dubious since salt is almost invariably present in commercial meat "emulsions". Tsai et al.[70] concluded from results of emulsifying capacity and stability, light microscopic examination, as well as heat gelation of proteins, that the functional properties of myosin are superior to those of actin or sarcoplasmic proteins. However, they noted that the emulsifying capacity of all the different protein fractions (myosin, actin, tropomyosin-troponin, and sarcoplasmic fractions) decreased with increasing protein concentration. At protein concentrations of 12 mg/mℓ or greater, the emulsifying capacities of all the proteins were equal. Since the protein concentration in a meat wiener system was estimated to be well above the critical point of 12 mg/mℓ, it was suggested that the value of using emulsifying capacity data obtained from dilute model systems in predicting functionality in real meat systems is questionable.

The increase in emulsifying capacity with decreasing protein concentration may be related to a greater degree of protein unfolding during the shearing involved in the emulsifying process, which may be aided by the hydrophobic association of the polypeptide chains with the droplets, resulting in a much larger, available volume/surface area of protein and thus enhancing emulsifying efficiency.[71,72]

The characteristics of meat proteins which affect their emulsifying properties are not well defined yet. The importance of soluble protein content on emulsifying capacity as well as other functional properties was widely emphasized for many years. For example, Carpenter and Saffle[73] studied various factors affecting emulsifying capacity of sausage meats and concluded that the amount of soluble protein determined emulsifying capacity in meat emulsions, and that factors affecting solubility also directly affected emulsifying capacity. Inklaar and Fortuin[74] also related emulsifying capacity to the amount of soluble protein and factors affecting myofibrillar protein solubility. Similarly, the emulsifying capacity of fish protein was related to changes in solubility induced by frozen storage, heating, or pH change.[75]

In general, the importance of soluble protein on emulsifying capacity has been well documented.[72] However, solubility by itself is not a good predictor of emulsifying and other functional properties. For example, Gillett et al.[76] discussed discrepancies in the relationship between soluble protein and emulsifying ability. They reported that soluble protein from fresh, uncooked, unfrozen meat sources was highly correlated with emulsifying ability, irrespective of the original meat source. For frozen or cooked meat, however, not only did the soluble protein content decrease, but the remaining noncoagulated soluble protein had a lower emulsifying ability and fell below the regression line for fresh meat sources; this observation was attributed to denaturation.

Hansen[62] also suggested that undenatured proteins are required for good emulsifying properties. He speculated that an excessive temperature-rise during chopping to give a final batter temperature of over 27°C, resulted in fat separation during smoking and cooking, probably due to partial protein denaturation and breaking of the protein matrix, giving rise to an unprotected fat dispersion. However, no data were actually presented to confirm that denaturation had indeed occurred at the high chopping temperature. Helmer and Saffle,[77] on the other hand, suggested that the breakdown of sausage emulsions at elevated chopping temperature was not due to denaturation of soluble protein, based on results from paper chromatography, electrophoresis, extractable protein, microscopy, and actual production. Perchonok and Regenstein[78] also suggested that even excessive chopping temperatures (30°C) failed to affect timed emulsification or emulsion stability, and that instability of commercial sausages is not due to changes in the protein caused by high chopping temperatures.

The relationship between the surface tension responses and emulsion stability was reported by Acton and Saffle.[79] Salt-soluble proteins from various meat sources exhibited surface tension responses corresponding to the type III response assigned to substances classified as surfactants.[80] Although both surface tension and emulsion stability were dependent on protein concentration, highly significant correlation coefficients were obtained for the relationships between surface tension values and stability ratings.

The importance of the concept of hydrophile-lipophile balance (HLB) on meat emulsion stability was emphasized by van Eerd.[81] The HLB system was originally devised by Griffin[82] to aid in the choice of surfactants or surfactant blends for maximum emulsion stability. In this system, HLB numbers are assigned to surfactants or emulsifiers depending on their hydrophobic or lipophilic and hydrophilic characters. HLB values may range from 1 (e.g., oleic acid) to 20 (e.g., potassium oleate). Emulsifiers with HLB values less than nine are considered to be hydrophobic, those with HLB values from 11 to 20 are hydrophilic, and those between 8 and 11 are intermediate.[60] Correspondingly, the "required HLB" number of an oil indicates the HLB number of the emulsifier which should be used to give maximum stability when an emulsion is prepared using that oil.

van Eerd[81] determined the required HLB values of various oils and fats to be approximately 14 for beef tallow, mutton tallow, lard, and castor oil, and approximately 10 for paraffin oil. The emulsion stability of mutton homogenates as well as protein fractions was investigated as a function of HLB of different blends of castor and paraffin oil. Emulsions prepared

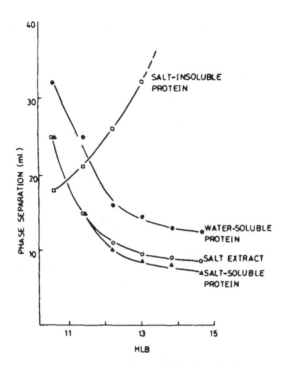

FIGURE 9. Effect of HLB on the stability of emulsions obtained using various mutton protein fractions. Emulsion stability was indicated by the extent of phase separation; a dashed line indicates that no stable emulsions were formed. (From van Eerd, J.-P., *J. Food Sci.*, 36, 1121, 1971. © Institute of Food Technologists. With permission.)

from prerigor meat homogenates in 3% NaCl solution were more stable than those from postrigor meat homogenates, and the HLB for optimum emulsion stability shifted from HLB >14.6 for prerigor meat to HLB =13 for postrigor meat, suggesting a more hydrophobic nature of the components in the postrigor system. Salt-soluble proteins and salt extracts containing both water- and salt-soluble proteins formed more stable emulsions than water-soluble proteins alone, with maximum stability at HLB >14.6; stability dropped off rapidly at HLB lower than 12. On the other hand, salt-insoluble proteins could form emulsions only at HLB lower than 13, with maximum stability at HLB lower than 10.6 (Figure 9). van Eerd[81] suggested that although the emulsifying properties of mutton were largely due to water- and salt-soluble proteins, the insoluble proteins also made a partial contribution to stability and, in addition, made the meat less sensitive to the influence of HLB.

The contribution of the insoluble protein fraction in meat emulsions has been suggested by various investigators. Swift et al.[63] noted that during emulsification, up to 84% of the proteins in solution became insoluble. Timed emulsification model systems were used to study the importance of low-salt and high-salt soluble and insoluble fractions of chicken breast muscle.[83,84] Emulsions formed using whole muscle or muscle washed with low molarity salt solution and suspended in buffered (10 mM sodium phosphate, pH 7.0) 0.6 M NaCl solution were more stable upon centrifugation than emulsions prepared using muscle in distilled water, or buffered 0.025 M or 0.05 M NaCl. The size of insoluble protein pellets after centrifugation of the emulsions decreased with increasing emulsification time, suggesting that constituents of the pellet phase were participating in emulsion formation and thus being progressively removed from the pellet phase upon prolonged emulsification.

Although high salt- (0.6 M NaCl) soluble proteins, i.e., the myofibrillar proteins, were the primary emulsifiers and emulsion stabilizers in the meat systems, good emulsifying properties could be observed in the fraction remaining after exhaustive washing of muscle to remove low and high salt-soluble fractions, by resuspension in 0.15 M or 0.6 M NaCl, pH 7.0. These studies thus raised the possibility of the contribution in emulsion formation of the insoluble protein fraction remaining after repeated salt extractions, which may contain both connective tissue and nonextractable myofibrillar proteins. The requirement of salt for maximization of the insoluble phase muscle properties was suggested to be related to the effect of salt on hydration state of the protein and viscosity, rather than on protein solubility, since it appeared that the proteins from the insoluble phase were directly incorporated into the emulsion without prior solubilization. It was speculated that while the initial content of high-salt extractable protein may give an indication of the total myofibrillar protein content present in a meat source as well as the ease and extent of tissue disintegration, it does not necessarily mean that the protein in a meat source must be soluble to participate in emulsions.[83]

Perchonok and Regenstein[85] also presented evidence to refute the common belief of many emulsion researchers that soluble proteins are necessary to form a stable meat emulsion. Table 11 shows the protein contents of the aqueous and pellet layers during timed emulsification studies using exhaustively washed chicken breast muscle. Most of the protein was either in the insoluble form (in the pellet) or in the emulsion, with only about 2% of the initial protein ever being in the aqueous layer, showing that in this case, salt-soluble proteins were not important participants in the emulsification process. After a 3-minute emulsification, the pellets contained only 6 to 13% of the protein compared to the 0-min sample pellets, indicating that during the emulsification, about 90% of the protein in the insoluble pellets had become part of the cream or emulsion layer. Heating of the emulsions before centrifugation showed no statistical differences for the emulsion stability or aqueous protein disappearance relative to temperature, suggesting that once formed, the emulsions were stable. On the other hand, if the heating step was conducted prior to emulsification and centrifugation, a great decrease in both emulsion stability and aqueous protein disappearance occurred between 60 and 75°C for the muscle. For emulsions prepared using natural actomyosin, the emulsion stability showed a decrease when heated above 40°C prior to emulsification, although the extent of aqueous protein disappearance did not decrease significantly until the actomyosin had been heated to between 60 and 75°C.

Borderias et al.[72] concluded that superior emulsifying properties of chicken muscle homogenate extract compared to fish homogenate arose chiefly from differing characteristics of the salt-soluble protein fractions, since the emulsifying properties of the water-soluble protein fractions were identical. They suggested that variations in weight of emulsified oil between 3 different species of fish (bonito, cod, and horse mackerel) were due mainly to the varying soluble protein concentrations, whereas differences between fish and chicken were due both to the varying soluble protein concentration as well as the differing behavior of the myofibrillar proteins. It was hypothesized that fish myofibrillar proteins may be inherently less stable, undergoing certain interactions which reduced their emulsifying ability, whereas the protein interactions were less intense in chicken muscle, resulting in effectively higher protein concentrations available for emulsification.

A linear relationship was found for apparent viscosity vs. protein concentration, as well as between apparent viscosity and amount of fat emulsified.[72] The apparent viscosity of chicken muscle extracts was higher than from fish, and appeared to be due to the myofibrillar protein fractions. Protein solubility of chicken and cod muscle proteins were lower than those from bonito and horse mackerel, an indication of possibly larger aggregates in the former, which may also have given rise to the observed higher viscosities. However, the possible molecular basis for the formation of larger aggregates in some species was not discussed.

Despite the seemingly obvious importance of hydrophobic interactions of muscle proteins

Table 11
PROTEIN CONTENTS OF THE AQUEOUS AND PELLET PHASES OF CENTRIFUGED EMULSIONS MADE WITH EXHAUSTIVELY WASHED CHICKEN BREAST MUSCLE

| Prepa-ration | Initial protein mg/mℓ | Emulsi-fication time (min) | Protein content at cooking temperature (°C)[a] | | | | | | | |
| | | | 0 | | 25 | | 50 | | 75 | |
			Aqueous (mg/mℓ)	Pellet (mg)	Aqueous (mg/mℓ)	Pellet (mg)	Aqueous (mg/mℓ)	Pellet (mg)	Aqueous (mg/mℓ)	Pellet (mg)
A	4.2	0	0.1	10.4	0.2	10.7	0.2	12.0	0.3	11.5
		3	0.2	1.1	0.2	1.4	0.2	0.7	0.2	1.3
B	5.1	0	0.2	13.6	0.2	13.0	0.2	13.4	0.4	13.3
		2	0.3	1.5	0.3	1.7	0.3	1.8	0.4	1.4
		3	0.3	1.4	0.3	1.4	0.3	1.2	0.4	1.4

[a]   Emulsions were heated after timed emulsification, before centrifugation.

Adapted from Perchonok, M. H. and Regenstein, J. M., *Meat Science*, 16, 31, 1986.

in influencing their behavior with fat particles or oil globules in emulsification, the question of the contribution of hydrophobic properties of muscle proteins to their emulsifying properties has been directly addressed only recently.[48,49]

Surface hydrophobicity of salt-extractable proteins from beef was determined using the fluorescent probe method.[48] Changes in protein solubility and hydrophobicity were monitored as a function of heating temperature, and related to changes in emulsifying properties determined as emulsifying activity index (EAI) and emulsifying capacity (EC). Table 12 shows the highly significant multiple regression models describing EAI and EC in terms of the solubility (s) and hydrophobicity (S_o) parameters. When samples of widely differing solubility were entered as data for the backward stepwise regression analyses, the parameter describing the interaction of s and S_o accounted for more than 71% of the variability in observed EAI values (see Table 12, Equation 1). Greater than 91% of the variability in the EC values could be explained by the three parameters s, $S_o \times s$ and s^2 (see Table 12, Equation 2), compared to a simple linear regression model describing EC only as a function of solubility, which could account for only 73% of the observed variability in EC.

Upon heat treatment, initial denaturation of proteins may occur with little or no apparent loss in solubility; the denaturation step is usually followed by aggregation and coagulation or gelation. Since the relationship between physicochemical properties of proteins and their functional properties may differ between soluble and insoluble systems, regression models were computed incorporating only samples with high solubility or only those with low solubility. As shown in Table 12, if only high solubility (s >50%) samples were considered, S_o alone accounted for over 70% and 82% of the variability in EAI and EC values, respectively (see Table 12, Equations 3 and 4), and solubility itself was not entered into the regression equations. When samples with low solubility (s <50%) were considered, solubility parameters were again incorporated into the regression equations; the squared terms S_o^2 and s^2 were significant in predicting EAI (see Table 12, Equation 5), while s^2 was significant in affecting EC (see Table 12, Equation 6). The statistical significance of the squared terms in these regression equations may be interpreted as curvilinear responses of the emulsion properties, initially increasing with increasing values of S_o and s, then reaching maximal emulsifying properties at intermediate S_o and s, and finally decreasing at high S_o and s values.

Table 12

PREDICTION EQUATIONS FOR EMULSIFYING ACTIVITY INDEX (EAI)
AND EMULSIFYING CAPACITY (EC) OF SALT-EXTRACTABLE BEEF
PROTEINS USING SURFACE HYDROPHOBICITY (S_0)
AND SOLUBILITY (s) PARAMETERS[a]

All samples

$$EAI = 0.0013S_0 \times s + 2.66 \qquad (n = 26, R^2 = 0.71, P<0.001)$$

$$EC = 6.46s + 0.0013S_0 \times s - 0.055s^2 - 35.3 \qquad (n = 26, R^2 = 0.92, P <0.001) \qquad (2)$$

High solubility samples (s $\geqslant$ 50%)

$$EAI = 0.0073S_0 + 3.49 \qquad (n = 8, R^2 = 0.70, P <0.010) \qquad (3)$$

$$EC = 0.10S_0 + 137 \qquad (n = 8, R^2 = 0.82, P <0.002) \qquad (4)$$

Low solubility samples (s < 50%)

$$EAI = 0.0000023S_0^2 + 0.0035s^2 + 1.42 \qquad (n = 18, R^2 = 0.63, P <0.001) \qquad (5)$$

$$EC = 0.094s^2 + 53 \qquad (n = 18, R^2 = 0.81, P <0.001) \qquad (6)$$

[a] From data of Li-Chan et al.[48] *J. Food Sci.*, 50, 1034, 1985.

These results showed that in general, both solubility and hydrophobicity were important parameters in describing emulsifying properties of salt-extractable proteins from beef muscle. S_0 was more important for influencing emulsification by samples with solubility over 50%, whereas solubility parameters were more influential than hydrophobicity for samples that had less than 50% solubility. These conclusions are also depicted in the three-dimensional surface plots showing the relationship between S_0, solubility and the emulsifying properties (Figue 10 A and B). Although there were slight differences in the plots for EAI and EC, it is apparent that in both cases, the interaction between the two parameters S_0 and s, rather than their individual values, is crucial for good emulsifying properties. Thus, Figure 10A shows that the highest EAI peaks resulted at high (85%) solubility and moderate (900) hydrophobicity, or at moderate (35 to 70%) solubility coupled with high (1200 to 1500) hydrophobicity. Figure 10B indicates that combinations of moderate to high values for solubility (50 to 100%) and hydrophobicity (1000 to 1500) favored high EC.

Contrary to popular belief of the detrimental effects of heating on protein denaturation and emulsifying properties, it was noted that moderate heating to 45°C or lower temperatures, often improved functionality of salt-extractable proteins from beef and fish, compared to unheated controls.[49] It was speculated that at temperatures below 50°C, the dissociation of muscle protein complexes and protein denaturation which have been reported to occur without accompanying aggregation[86-88] were reflected in the increased surface or effective hydrophobicity of the salt-extractable proteins, with little change in solubility properties or total sulfhydryl group content. Under these moderate heating conditions, emulsifying properties as well as fat-binding properties were often improved. However, at higher temperatures, functionality was impaired, probably due to aggregation as reflected by the lower solubility and sulfhydryl content, and to some extent at temperatures above 70°C, also by decreasing hydrophobicity. Thus, conditions favoring an optimum balance of high solubility, hydrophobicity and sulfhydryl group properties were considered to be necessary for good emulsifying properties. Net surface charge on the salt-extractable proteins was considered to have only minimal effects on the emulsifying properties, due to the presence of high (0.6 *M*) salts during emulsification.

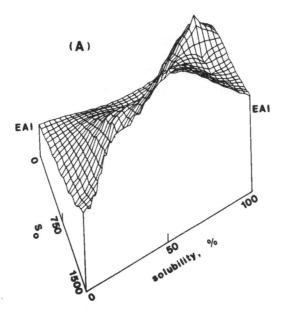

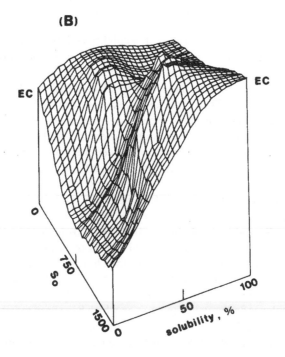

FIGURE 10.   Three-dimensional surface plots depicting the relationship between hydrophobicity (S_o), solubility, and emulsifying properties [(A) EAI and (B) EC] of salt extracts from beef top round. (From Li-Chan, E., Nakai, S., and Wood, D. F., *J. Food Sci.*, 49, 345, 1984. With permission.)

B. Water- and Fat-Binding Capacities

The ability of proteins to bind water and fat, as well as to retain these two components upon heating and storage, is crucial in the manufacture of processed meat products. These water- and fat-binding properties determine not only the final cook yield, but also the appearance (e.g., absence of fat cap and jelly formation), and textural and sensory characteristics (e.g., juiciness and flavor release) which constitute final product quality.

Water absorption, water binding and water holding are terms which have been used somewhat interchangeably in the literature. Water binding and water holding usually refer to the water which is retained by the sample following such treatments as filter paper press, centrifugation or gravity filtration.[89-92] The terms "absorbed" moisture of brine and "retained" moisture or brine were coined by Porteous and Wood[93] to differentiate between the water-binding parameters determined in uncooked and cooked samples, respectively. "Swelling" properties have also been suggested as a measure of water absorption.[94] Water absorption or adsorption usually refer to water taken up spontaneously by a dry protein powder after equilibration against water vapor of a known relative humidity.[71]

The water of skeletal muscle exists in three forms which can be designated as bound, restricted, and free or bulk water.[59] The bound water, estimated at 8 to 10% of the muscle water is approximately equal to the first and second layers of water bound to protein hydrophilic groups. The restricted water phase is considered the fraction typically measured by water-holding capacity tests, and can vary, depending on factors such as salt addition, pH, and tissue disruption. The free or bulk water is held by capillary action and is easily expressible by pressure. Water or ice added during comminution initially is in the bulk water form, but may become part of the restricted water form by salt-induced myofibrillar protein changes.

Unlike emulsification properties which involve the blending together of liquid fat and aqueous protein solutions, fat-absorption or fat-binding capacity is usually measured by adding excess liquid fat or oil to a dry protein powder, thoroughly mixing and holding, then determining the amount of bound oil after centrifugation.[95-97] By this methodology, the fat absorption has been attributed mostly to physical entrapment of the oil, and fat-absorption capacity could be correlated with increasing bulk density of the protein sample.[71] A turbidimetric method has also been developed to measure fat-binding capacity.[98] After the initial centrifugation of the oil and protein powder, the protein pellet was repeatedly washed with water and then reprecipitated with metaphosphoric acid to remove loosely bound oil physically adsorbed on the surface of the protein pellet, thus minimizing the contribution of entrapped oil to the true fat-binding capacity determination.

Despite the growing acceptance of the inaccuracy of describing comminuted meat products as emulsions, and despite the growing awareness of the unique behavior of fat dispersed in protein gel matrices,[10] few studies have been undertaken to study the fat-binding capacity of muscle proteins, probably due to the lack of simple or established methodology for measuring true fat-binding capacity. Thus in most studies, the term "emulsifying capacity" has been used synonymously for "fat-binding capacity" of meat systems. However, as Meyer et al.[99] reported, commercial food emulsifiers added to sausage batter in fact decreased the fat-binding of the batter system.

Effects of heating, freezing, and pH on both fat-binding and emulsifying properties of salt-extractable proteins from beef top round and rockfish fillets were studied in relationship to corresponding changes in the protein dispersibility, surface hydrophobicity, and sulfhydryl group content.[49] Emulsifying and fat-binding properties generally followed parallel trends, being improved under conditions favoring balance of high dispersibility, hydrophobicity, and sulfhydryl content, which corresponded to samples which were moderately heated (<45°C), at higher pH (7.0 vs. 5.5) and were in the fresh rather than the frozen form. However, unique properties were noted in the fat-binding properties of fish salt-extractable proteins.

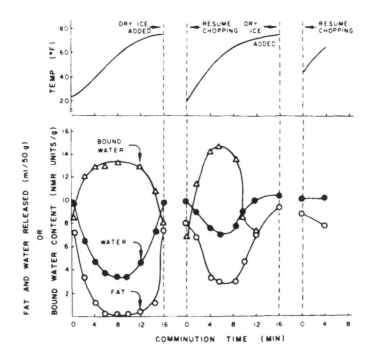

FIGURE 11.   Effect of chopping and two cooling cycles on binding of fat and water and the batter temperature. (From Brown, D. D. and Toledo, R. T., *J. Food Sci.*, 40, 1061, 1975. © Institute of Food Technologists. 19, 247. With permission.)

High temperature-low pH treatment of salt extracts from fish favored gelation and simultaneously significantly increased fat binding while decreasing emulsifying properties. Under these conditions, the fish proteins had high hydrophobicity and low dispersibility, and it was speculated that increased hydrophobic interactions at high temperature and low pH might be important in both the enhanced gel formation and fat-binding capacity.

The interrelationship between water binding and fat binding of meat has been suggested.[100] A close parallel between water and fat binding appears during the comminution step in sausage preparation (Figure 11). After a comminution time of 6 to 14 min, both fat and water were concurrently bound to a maximal level, as reflected by the minima in release of these two components as well as the maximum in the bound water level measured by nuclear magnetic resonance.[100] This stage was also accompanied by an increase in temperature to between 18 and 21°C. Further, comminution resulting in even higher temperatures led to rapid declines in both fat- and water-binding abilities. During the initial phase of comminution, the batters appeared to lose their binding capacity for water earlier than for fat, suggesting that as water was released, the ordered structure of protein, water, and fat which was responsible for mechanical entrapment of the rest of the fat particles had disintegrated, resulting in fat release. Brown and Toledo[100] suggested that after the first comminution, the protein forming the ordered structure responsible for the ability for fat and water binding had been partially damaged irreversibly; on cooling and subsequent rechopping, the minima in the fat- and water-release curves were not at the same low level as during the first comminution. It was also shown that in some formulations, there was an inverse relationship between fat and water binding, suggesting a competition between the two components for binding in the muscle system.

Most of the factors which increase the water-holding capacity of meat have also been suggested to improve the distribution of fat in sausage products. It was hypothesized that fat retention may be governed by similar factors responsible for water holding or swelling capacity of the mince of lean beef.[101] Factors which affect the water-holding capacity of meats have been extensively studied, as reviewed by Hamm.[89,101] It is generally accepted that water-holding capacity is increased by the use of prerigor vs. postrigor meat, fresh instead of frozen meat, by conditioning or aging, by higher pH, and by incorporation of salts and polyphosphates. However, the analysis of the mechanism of these phenomena is a complex problem, making difficult the prediction of ability to bind water and fat, or conversely, the release of water and fat upon storage or cooking of the product.

For example, Swift and Berman[102] suggested that a higher moisture-to-protein ratio as well as a higher fat content were associated with increasing water retention. In contrast, Miller et al.[91] reported decreasing water-holding capacity with increasing fat and increasing moisture-to-protein ratio; they suggested that the relationship between the moisture-to-protein ratio and moisture-loss determined by centrifuge and cookout could be used to predict moisture-loss in sausage products. Mittal and Blaisdell[103] suggested a two-term model to predict weight loss in frankfurters during thermal processing, based on the moisture-loss rate being proportional to product temperature and inversely proportional to fat-protein ratio.

The importance of solubility of myofibrillar proteins to water holding capacity has also been controversial. Torgerson and Toledo[104] found a significant correlation between fat release upon cooking and the solubility of protein additives, including low-temperature extracted chicken meat protein, peanut protein, whey protein concentrate, and single cell protein. The water-absorption capacity of the chicken protein was high, while the solubility of this protein in water was low. Upon increasing temperature, the water solubility increased while water absorption capacity decreased.

The concurrent loss of myofibrillar protein solubility and water-holding capacity during rigor mortis has been interpreted by many researchers as an indication of the importance of solubilisation of the actin-myosin system for fat and water binding. However, the inaccuracy of this concept was pointed out by Hamm.[101] In prerigor salted beef, for example, although solubility of the myofibrillar proteins is as low as that found in the rigor state, the water-holding capacity of prerigor salted beef and the quality of sausages prepared from it are much higher than the corresponding properties in the rigor samples, and in fact are comparable to sausages made from hot prerigor beef with high myofibrillar protein solubility. In addition, the effect of diphosphate on the water-holding capacity of raw and heated, salted muscle homogenate cannot simply be related to its effects on increased protein solubilisation. Although moderate diphosphate addition resulted in strong swelling and increased water-holding capacity, higher levels of diphosphate, which promoted further protein solubilisation, was detrimental to the water-holding capacity.

It has been estimated that the amount of water bound to the surface of proteins is only about 0.5 g/g protein, thus yielding only about 0.1g/g meat for meat with 20% protein content. It has thus been concluded by some investigators that the binding of water to the surface of protein molecules is too small to account for the observed changes in water-holding capacity.[89] Kinsella[71] noted that prediction of protein hydration and water binding has been attempted based on amino acid composition, but disparity between researchers regarding the extent of hydration of different polar and nonpolar amino acid residues has been prevalent.[4,71]

Offer and Trinick[105] suggested, from studies on myofibrils, that the contribution of water bound to protein surfaces is only a small fraction of the total water present. Instead, they concluded that most of the water is held by capillary forces between the thin and thick filaments. They hypothesized that interfilament spacing was the major determinant of the water-holding capacity of myofibrils, and that this spacing depends mainly on long-range electrostatic forces.

Although the bound water content on the surface of protein molecules may be small, the importance of protein-water interactions to explain the water-holding capacity of meat should not be discounted. Muscle proteins are capable of forming a gel-like network which can entrap large amounts of water. It has been suggested that salt addition would cause solubilisation of myosin and the extracted myosin would probably form a gel or matrix, resulting in the enhanced water uptake. Ashgar et al.[4] pointed out that especially in the case of salt-induced swelling, the neutral salts influence structural stability of the muscle filaments not only by electrostatic effects of ions, but also by the lyotropic or Hofmeister effect which originates from other properties such as polarizability and dipole moments, which may be independent of charge. It is likely that salt incorporation may intensify protein network formation via increased hydrophobic interactions and reduced electrostatic repulsion, thus increasing both the water- and fat-binding properties.

The importance of hydrophobic interactions in the water- and fat-binding properties of muscle proteins was demonstrated using fluorescence probe based measurements of hydrophobicity. Changes in both content and surface hydrophobicity of salt-extractable protein were observed to accompany the loss of water-holding capacity of postrigor beef neck muscle, in comparison with prerigor muscle.[56] Decrease in the retained moisture (as water or brine) was observed at 3 and 10 days postslaughter, compared to 0 days (prerigor) and 1 day postslaughter. Salt-extractable content was highest at 0 days, decreased at 3 days, then increased again slightly at 10 and 14 days. Surface hydrophobicity of the proteins showed a gradual increase over time, being significantly higher at 10 days.

Changes in functional properties, including water binding, fat binding, emulsifying, and gelling properties were investigated in chicken breast muscle as a function of postmortem holding time prior to deboning.[50] Generally, muscle deboned at 2 days had inferior functional properties, compared to muscle deboned either at 0 or at 7 days. Salt-extractable proteins from 2-day-old muscle did not differ in content, solubility, hydrophobicity, or total sulfhydryl groups compared to 0-day-old muscle. However, subtle changes in the 2-day-old proteins were suggested by their resistance to mild heating-induced changes in the protein structure, as reflected by their lower hydrophobicity after heating at 40°C, compared to 0- and 7-day-old extracts heated at the same temperature (Figure 12). On the other hand, the content of salt-extractable proteins from 7-day-old muscle was higher than at 0- and 2-day-old muscle. These 7-day-old proteins had lower solubility and higher hydrophobicity, supporting the suggestion by Johnson and Bowers,[106] that higher molecular weight components or complexes were extracted from aged muscle. Upon heating at high temperature (80°C), the 7-day-old proteins showed a hydrophobicity increase less than the correspondingly heated proteins from 0- and 2-day-old samples (Figure 13).

Fat binding by myofibrillar proteins from fish muscle has been investigated to a large extent due to the possible participation of lipid-protein interactions in denaturation of fish proteins, especially upon frozen storage.[107-110] Deterioration of fish-muscle protein properties has been suggested to be a result of free fatty acids and/or lipid peroxides; however the question of whether interactions between proteins and neutral and polar lipids are detrimental or beneficial to fish protein stability, remains controversial. Some evidence exists, especially in biological systems, to suggest the essential and protective role of intact lipids on stability of fish myofibrillar proteins. It has also been proposed that the neutral lipids may diminish or counteract the detrimental effect of free fatty acids either by dissolving them and diluting out their effective concentration, or by actually competing with them for binding sites on the protein. In contrast, other studies, particularly in model systems, suggest the detrimental effects of lipid-protein complex formation which denature the protein and deteriorate its functional properties. In addition, when denaturation of fish proteins was triggered by other factors such as heating or foam formation, the degree of denaturation has been frequently intensified in the presence of fish lipids. Model studies have indicated that when lipids and proteins extracted from the same fish were incubated, large insoluble protein-lipid complexes

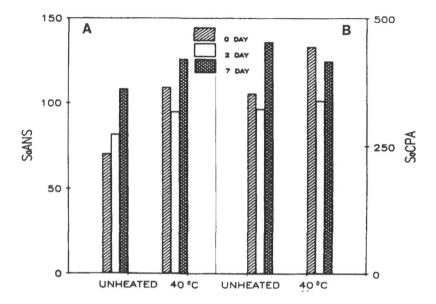

FIGURE 12.   Changes in hydrophobicity [(A)S₀ANS and (B)S₀CPA] of proteins in unheated and 40°C-heated salt extracts from chicken breast muscle at 0, 2, and 7 days postmortem time. (From Li-Chan, E., Kwan, L., Nakai, S., et al., *Can. Inst. Food Sci. Technol. J.*, 1986, 19, 247. With permission.)

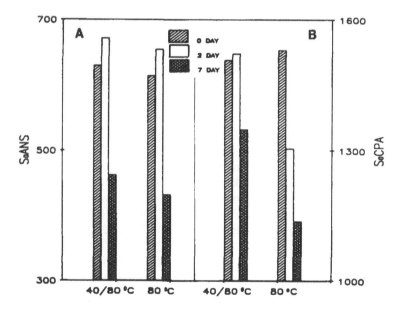

FIGURE 13.   Changes in hydrophobicity [(A)S₀ANS and (B) S₀CPA] of proteins in heated salt extracts from chicken breast muscle at 0, 2, and 7 days post mortem time. Extracts were heated at 80°C for 15 min, or at 40 and 80°C each for 15 min. (From Li-Chan, E., Kwan, L., Nakai, S., et al., *Can. Inst. Food Sci. Technol. J.*, 1986, in press. With permission.)

Table 13
**EFFECT OF VARIOUS TREATMENTS ON
PERCENTAGE OF LIPID BOUND TO ACTIN**

Treatment	(%) lipid bound to actin			
	Monomer	Polymer	Pellet	Total
Neutral lipids				
At 4°C	10	23	0	33
At room temp.	13	33	0	46
With Ca^{++}	3	0	90	93
With Mg^{++}	21	6	62	89
Heated	16	25	8	49
Agitated	12	31	20	63
Polar lipids				
At 4°C	—	48	0	48
At room temp.	—	57	0	57
With Ca^{++}	—	8	31	39
With Mg^{++}	—	12	21	33
Heated	—	13	13	26
Agitated	—	20	12	32

Note: ^{14}C-lipids bound to monomer, soluble polymer and pellet
forms of actin were quantitated following sucrose gradient
centrifugation; bound lipid was calculated as a percentage of
the added lipid (400 μg lipid added per 7 mg actin).

Adapted from Shenouda, S. Y. K. and Pigott, G. M., *J. Food Sci.*,
40, 523, 1975.

were formed, which had quite distinct properties from naturally occurring lipoproteins.
Electron paramagnetic resonance studies as well as observations of the resistance of the
aggregates to dissolution or dissociation in urea and SDS suggested the existence of strong
hydrophobic participation.[108]

Pure fish actin preparations interact easily with fish lipids.[111,112] As Table 13 shows, upon
incubation at 4°C, actin bound 33 and 48% of the added neutral and polar lipids, respectively.
It was suggested that hydrophobic bonding was involved either in binding part of the lipid
to the protein, or in holding the protein molecule in a specific configuration to enclose the
lipids. The higher binding of polar rather than neutral lipid was speculated to be due to
possible ionic interactions between charged groups, as well as to the greater degree of
unsaturation of fish polar lipids than neutral lipids, which would favor stronger hydrophobic
interactions.

The amount of total lipid bound to actin increased at higher incubation temperature, in
the presence of Ca^{2+} or Mg^{2+} cations, and by heating or agitation. However, the effects
on polar and neutral lipids were not always parallel (Table 13). Shenouda and Pigott[111]
concluded that fish actin interacts with polar as well as neutral lipids at cold and room
temperature, but the F-actin form interacts 2 to 3 times more strongly than the G-actin form.
Treatments inducing G- to F-actin transformation increased lipid-actin complex formation.
Increasing the incubation temperature from 4°C to room temperature, induced conformational
changes resulting in greater binding of both polar and neutral lipids. However, changes
resulting from Ca^{2+} or Mg^{2+} induced polymerization, heating or agitation probably caused
exposure of more hydrophobic regions which increased neutral lipid binding while decreasing
polar lipid binding. Participation of hydrophobic interaction was indicated by the actin-lipid
interaction even at high ionic strength. When the ionic strength was increased over 1.0,

a steeper increase in neutral lipid binding was observed, indicating salt-induced changes in the actin molecules pertaining to a more hydrophobic nature.[108] Electron paramagnetic resonance studies showed that the hydrophobic interactions between neutral lipids and fish actin were stronger than the forces binding polar lipids to actin.[113] However, electrostatic interaction was also involved in lipid binding, especially in the case of actin-polar lipid interaction. The stability of the complexes to SDS and urea treatments was suggested to indicate either existence of covalent binding, or unusually strong electrostatic and/or hydrophobic bonding involved in the complex formation.[111]

Unlike the ease of interaction of actin with lipid, no interaction occurred between freshly prepared, undenatured myosin, and polar or neutral lipids.[112,114] However, myosin-lipid complex formation could be induced by treatments including aging, agitation by foam formation, or heating of the myosin preparation. Polar lipids showed more tendency than neutral lipids to interact with heat-denatured myosin, while addition of the nonionic detergent Triton X-100® to myosin significantly reduced the amount of neutral lipid bound during heating. Once formed, the lipid-protein complexes were difficult to separate again, even by severe physical means or exhaustive washing. Strong hydrophobic and ionic interactions were implicated in forming the lipid-myosin complexes.[112]

C. Gelation

The establishment of a stable gel network is closely linked to various functional properties of muscle-based foods, including water-, fat- and particle-binding and texturization.[50] The setting of a meat emulsion in comminuted products as well as the binding of chunks of meat in restructured or reformed products are believed to be based on the formation of a stable protein gel. Generally, gelation is considered to be a heat-initiated process since raw meat pieces do not exhibit significant binding to each other.[4] The exception to this observation is the low temperature-setting of sols from certain fish species, which can occur upon mild heating at 40°C or even upon cold storage at 4°C.[115,116]

The changes in myofibrillar protein structure by increasing temperature have been the subject of many investigations.[59,86-88,117-123] Ziegler and Acton[87] summarized the following changes in natural actomyosin which occur with increasing temperature. At 30 to 50°C, native tropomyosin is dissociated from the F-actin superhelix, which itself is dissociated into single chains at 38°C. At 40°C, myosin light chains dissociate from heavy chains, followed by conformational changes in the head and hinge regions of the myosin molecules. The actin-myosin complex dissociates at 45 to 50°C, followed by a helix-coil transformation in the light meromyosin chains at 50 to 55°C. At this stage there is rapid aggregation of protein molecules. Finally, at temperatures above 70°C, major conformational changes occur in the G-actin monomer.

Unfolding of the protein molecules upon thermal treatment has been confirmed by various investigators. For example, an increase in the exposed sulfhydryl groups as measured by titration with sulfhydryl group reagents (PCMB, NEM) has been observed.[117] Heating of muscle tissue results in increase of pH, starting at 30°C, with maximum pH increase at between 40 and 60°C. The resulting shift of the isoelectric point of myofibrillar proteins to higher values (e.g., from 5.0 to 6.0 after heating at 80°C) have been attributed to conformational changes and unfolding which exposes previously masked basic groups, including some imidazolium groups of histidine.[124] Transfer of aromatic amino acid residues, especially tryptophan, from a hydrophobic environment to a more exposed or polar environment was indicated by Samejima et al.[88,123] and Ishioroshi et al.[125] using fluorescence and difference spectra, optical rotatory dispersion, and circular dichroism measurements. Li-Chan et al.[48,49] showed that surface hydrophobicity of the salt-extractable proteins increased upon heating.

From these results, it has been suggested that a change in properties, including solubility of myofibrillar proteins between 30 and 60°C is accompanied by an unfolding of the protein chain, followed at higher temperatures by an association of unfolded protein molecules which may lead either to formation of a coagulum or a gel network. The nature of the association links formed between the protein molecules at temperatures above 50°C is still not well defined. Hydrophobic groups were suggested to play an important role.[49,117,118,126,127] At temperatures below 70°C, sulfhydryl reactivity is not typically involved in the aggregation. However, at temperatures above 70°C, the total sulfhydyl group content decreases, presumably by oxidation to disulfide groups, whereas at even higher temperatures (over 80°C), a loss of total sulfhydryl plus disulfide groups occurs by oxidation to cysteic acid or splitting-off of hydrogen sulfide.

Samejima et al.[88,123] proposed that aggregation of globular head segments of myosin molecules was associated with sulfhydryl reactivity and oxidation to disulfide groups, whereas gel network formation was by association of unfolded helical tail segments. The bonding of the tail segments was suggested to involve noncovalent interactions such as hydrogen bonds and hydrophobic interactions, since gelation was inhibited by urea or guanidine hydrochloride. Liu et al.[126] also confirmed the importance of hydrophobic interactions in Atlantic croaker actomyosin gels, which could be solubilized with 2% sodium dodecylsulfate in the absence of sulfhydryl-reducing agents such as dithiothreitol. They believed that although oxidation of sulfhydryl groups could further facilitate the aggregation of the protein molecules, hydrophobic interactions were the predominant force bringing about aggregation and coagulation.

The role of individual muscle proteins, as well as myosin subfragments and subunits in binding and gelation phenomena, has been the center of many investigations,[3,87,88,125,128-139] as well as the subject of a recent review.[4] The precise contribution of each component of the myofibrillar protein complex to development of good gel properties is still not agreed upon by various researchers. However, it is generally accepted that the predominant influence on gelation in comminuted meat products is related to the physicochemical properties of myosin and/or actomyosin.

Factors affecting heat-induced gelation of myosin have been investigated.[4] Generally, the heat-induced gel strength of myosin has been found to increase with the square of the protein concentration; the highest shear modulus was obtained with KCl concentration of 0.1 to 0.2 M for myosin, whereas for actomyosin one rigidity peak appeared at about 0.2 M KCl and the other at 0.6 M KCl; the optimum pH of gelation has been suggested to be around 6.0, but this pH optimum varied with ionic strength and protein concentration; a heating temperature range of 55 to 60°C has yielded high gel rigidity, but again this varied with ionic strength and pH conditions and protein composition and concentration. Other factors affecting gelation include the age of the protein, since myosin stored in 0.6 M KCl was unstable and started losing its gel-forming potential after storage, which appeared to be correlated with an increase in viscosity, suggestive of aggregate formation.

Myosins from different species, as well as different isoforms of myosin even from the same species, have been shown to possess different heat-induced gelling properties; however the source of these differences is still not clear. Ashgar et al.[24] and Samejima et al.[3] noted differences between gelling of cardiac myosin, and red and white skeletal muscles. Cardiac muscles appeared to form much stronger gels than skeletal muscle myosin, whereas white myosin exhibited greater gel strength than red myosin. Differences in red myosin were also observed to be influenced by nutritional stress, the gel strength of myosin from underfed broilers being about 50% less than that of controls. Samejima et al.[3,139] suggested that the gel strength potential of different myosins depends on their heavy chains although the light chains may play some role of providing stability to the myosin gel.

Actomyosins, myosins, and muscle sols from fish, beef, pork, turkey, and chicken muscles have been reported to differ greatly in their gelling properties.[49,50,52,53,126,131,140-143] Some fish species are also unique in their ability to set at low (4°C) or moderate (40°C) temperature.[40,41,115,116,144-151]

For example, the extent of protein-protein interaction as measured by changes in light scattering was greater for beef actomyosin than mackerel actomyosin below 40°C; at higher temperatures, the converse was true.[130] Beef actomyosin solutions tended to precipitate upon aggregation with prolonged heating whereas the mackerel actomyosin aggregates remained in solution. Li-Chan et al.[49] also noted that heating at pH 5.5 resulted in stronger gel formation of salt-extractable proteins from rockfish, whereas coagulation or weak gel formation resulted from the salt-extractable proteins from beef. The fish proteins had greater aromatic and aliphatic hydrophobicity values than the beef proteins. Atlantic croaker and chicken actomyosins also showed differences in thermal denaturation and aggregation.[126] Despite their similar amino acid composition, the croaker and chicken actomyosins displayed differing viscosities and the croaker actomyosin underwent a greater degree of protein unfolding and associated changes in the shape and/or specific volume of protein molecules with increasing temperature than the chicken actomyosin.

Continuous evaluation during heating of sols from beef, fish, pork, and turkey muscles indicated the same general patterns for modulus of rigidity and energy loss for all the samples, but the major transitions occurred at lower temperatures and yielded a higher modulus of rigidity for fish (surimi) than other species.[140] The apparent energy loss vs. internal temperature plots showed least elastic character for the beef gels while turkey gels had nearly perfect elastic character after heating at 90°C. For surimi, the rapid increase in elasticity occurred at about 35 to 40°C, near the temperature of the initial rigidity peak, and it was suggested that preincubation at this temperature favored the contribution of hydrophobic interactions, which was crucial in producing stronger and more elastic gels than fish gels heated directly at the higher temperature of 65°C.

The importance of hydrophobic interactions in influencing the ease of setting at around 40°C in fish species was suggested by Niwa and coworkers in a series of investigations on fish and mammalian species,[40-47,144-146,152] and also by the induction of setting by introduction of aromatic or aliphatic groups on the muscle proteins.[45-47] Wu et al.[115,116] also supported the setting at 40°C as resulting from unfolding by thermal denaturation of particular regions of myosin, with subsequent formation of a network structure through aggregation of unfolded molecules. On the other hand, a low-temperature (4°C) setting did not show any discernible changes in differential scanning calorimetry thermograms, indicating different changes of fish protein during low-temperature setting from those during high-temperature setting. The physicochemical nature of this phenomenon is not known at present,[115] but the increase in hydrophobicity upon aging of myosin noted by others may be influencing the setting phenomenon.

The effects of fatty acid salts and analogs on the thermal stability and gel formability of myosin were reported by Egelansdal et al.[153] Increasing aliphatic chain length decreased thermal stability. All the detergents studied enhanced gel strength, especially potassium decanoate and dodecanoate; in the case of potassium dodecanoate and sodium dodecylsulfate, the enhancement was detectable only after a prolonged incubation time. It was suggested that an increase in gel strength by binding of the amphiphiles was most likely due to a reduction in random aggregation arising from increased repulsion of the protein chains with the fatty acid anions, which would favor a more ordered aggregation (gelation) upon subsequent heat treatment. Although not explored by Egelansdal et al.,[153] it is likely that enhanced hydrophobic interactions arising from the introduction of the hydrophobic aliphatic side chains on the myosin molecules could also contribute to the greater gel strength upon heating.

The addition of an aliphatic alcohol to fish myosin also had an effect on gel formation.[154] The phenomenon of "himidori" or thermally induced disintegration of oval filefish myosin gel was eliminated by the addition of n-butyl alcohol. It was suggested that the aliphatic alcohol modified the protein structure from an easily disintegrated type into one possessing a strong gel network, thus emphasizing again the role of the hydrophobic aliphatic side chains in gelation of fish myosin.

VI. FUNCTIONAL PROPERTIES AND MUSCLE PROTEIN STRUCTURE — QSAR APPROACH

In the past two decades, a lot of focus has been placed on functional properties of muscle proteins, particularly the salt-soluble proteins. Subsequent to the report by Hansen[62] on emulsion formation in comminuted meat systems and the investigations of variables influencing the determination of emulsifying capacity (EC),[63,73,155] equations have been suggested for quantitative incorporation of EC into the concept of binding properties of proteins for the purpose of use in least cost formulation computer programs.[9,156] Subsequent modifications of the original bind-constant system have been proposed to better define the functional properties of ingredients used in sausage products.[7,8,157,158]

However, in recent years, it is becoming gradually more obvious that the parameters commonly incorporated into bind constants, i.e., emulsifying capacity and protein or soluble protein content, are not sufficiently accurate indicators of final product quality. Gillett et al.[76] noted that while bind constants have proven useful in computer programs for least cost analysis for ingredient formulations, approximately 30% of the predictions are in error on emulsion stability.[159] These errors can result either in poor product quality due to underestimation of the requirement for functional ingredients, or in needlessly high product cost due to overestimation of the need for expensive ingredients.

Comer and Dempster[12] reported that emulsifying capacities were a poor basis for a comprehensive bind value scale and proposed instead an alternative approach based on the functional effects of the ingredients. They developed a hypothetical bind value scale based on total protein in the ingredients and on arbitrary protein quality factors, the latter being subjectively determined. For further work, they recommended the development of a functional property test reflecting gelation properties of ingredients, followed by the practice of "refinement of bind values by practical experience" to obtain values suitable for actual formulation and processing conditions. The need for new concepts in estimating contribution of meat and nonmeat ingredients, especially with emphasis on water-holding capacity and gelation phenomena, was also pointed out by Parks et al.[8] and the challenge to apply these techniques to develop a better data base for least-cost formulations was emphasized by Regenstein.[160]

In view of the necessity for accurate assessment of ingredient functionality for incorporation into an improved least-cost formulation program for blending of ingredients in comminuted product manufacture, there is an urgent need for quantitative information on the relationship between protein physicochemical or structural properties and their functional properties. Using a quantitative structure-activity relationship (QSAR) approach, equations could be developed to estimate the functionality contribution of available ingredients to comminuted meat products, by screening through data banks containing information on their basic physicochemical properties. Functionality values obtained by prediction from basic physicochemical properties of the ingredients may be more reliable compared to values obtained by direct measurement of functional properties, due to the easier standardization and greater reproducibility of methodology to measure basic physicochemical properties. In contrast, methodology for functional property determination is hardly standardized between researchers or meat processors, and functionality data are fraught with inconsistencies and high deviations resulting from differences in instruments and conditions for determination. An additional advantage to using predicted versus measured functional data is that a single set of data on basic physicochemical properties can be used to predict diverse functional properties such as gelation, cook loss, water binding, fat binding and emulsifying capacity.

Table 14
QUADRATIC EQUATIONS FOR PREDICTING FAT-BINDING AND EMULSIFYING PROPERTIES OF SALT-EXTRACTABLE PROTEINS FROM BEEF TOP ROUND

Emulsifying Capacity EC = $18.8D + 1.40CPA + 24.4SH - 0.049D^2 - 0.0099D*ANS - 0.15D*SH - 0.021CPA*SH - 1578$

$$(n = 74, R^2 = 0.77, P < 0.001)$$

Fat-Binding Capacity FBC = $-0.15ANS - 0.0079CPA + 0.000057D*ANS - 0.000019ANS*CPA + 7.76$
$$(n = 52, R^2 = 0.66, P < 0.001)$$

Emulsifying-Activity Index EAI = $0.58D - 0.0037D*SH - 2.95$
$$(n = 52, R^2 = 0.83, P < 0.001)$$

Emulsifying-Stability Index ESI = $0.027CPA + 0.30SH - 0.0017D^2 - 0.000051ANS^2 - 0.0000065CPA^2 - 0.000074D*CPA - 0.0029D*SH + 0.036ANS*SH - 0.00028CPA*SH - 21.8$
$$(n = 52, R^2 = 0.74, P < 0.001)$$

Note: Quadratic equations were computed by backward stepwise multiple regression; D = dispersibility; ANS = hydrophobicity determined using ANS; CPA = hydrophobicity determined using CPA; SH = sulfhydryl content. Samples included unheated and heated (35 to 75°C) extracts.

Adapted from Li-Chan, E., Nakai, S., and Wood, D. F.[49] *J. Food Sci.*, 50, 1034, 1985.

Thus, data banks containing information on basic properties allow for a more versatile search for suitable ingredients in a blend, depending on the desired functional performance in the product.

Early work investigating the quantitative relationship between physicochemical parameters and the emulsifying properties of salt-extractable proteins from beef showed that the emulsifying properties could be described more accurately by multiple regression equations incorporating solubility and hydrophobicity parameters (Table 12 and Figure 10), than by the simple relationship between emulsifying properties and solubility only.[48] Subsequently, two sets of multiple regression equations were computed for salt-extractable proteins from beef and rockfish, respectively, including samples which had been variously treated by heating, freezing, or pH adjustment.[49] For both beef and fish, it was demonstrated that highly significant equations for the prediction of fat-binding and -emulsifying properties were obtained when dispersibility, hydrophobicity, and sulfhydryl parameters were used as the independent or predictor variables (Tables 14 and 15). The equations show that emulsifying properties of both beef and rockfish were generally better in samples with an optimum balance of high dispersibility, hydrophobicity, and sulfhydryl groups. Fat-binding properties of rockfish proteins were favored under conditions of high temperature-low pH heating which resulted in lower dispersibility and higher hydrophobicity of the salt-extractable proteins. Under those conditions, the rockfish extracts also formed gels. Greater involvement of hydrophobic interactions under these conditions which favored gelation and fat binding were suggested for rockfish proteins, compared to beef proteins which only formed weak gels or coagulated, and also showed decreased fat binding under similar conditions of heating.

Muscle proteins from different species can vary widely in their functional and physicochemical properties (Tables 9 and 10). Although variations within species may be more accurately and precisely predicted by applying the QSAR analysis to each species (e.g., Tables 14 and 15 for beef and rockfish, respectively), this approach may limit the potential for versatility in choosing formulation ingredients from alternative sources of muscle products. Thus, general muscle protein structure-function relationships and discrimination of functionality were investigated by multivariate analyses of data pooled from a multitude of different meat ingredients.[53] Since this study probably represents the first application of

Table 15
QUADRATIC EQUATIONS FOR PREDICTING FAT-BINDING AND EMULSIFYING PROPERTIES OF SALT-EXTRACTABLE PROTEINS FROM ROCKFISH

Emulsifying Capacity EC = 0.43CPA + 3.79D + 4.94SH $-$ 0.00026CPA^2 $-$ 0.0054ANS*SH $-$ 0.036D*SH $-$ 138

$$(n = 70, R^2 = 0.81, P <0.001)$$

Fat-Binding Capacity FBC = -0.013ANS $-$ 0.22D^2 $-$ 0.000026ANS*CPA + 0.00017ANS*D $-$ 0.000075ANS*SH + 0.00015CPA*SH + 6.47

$$(n = 70, R^2 = 0.76, P <0.001)$$

Emulsifying-Activity Index EAI = 0.085CPA + 0.53D $-$ 0.000083ANS*CPA $-$ 0.00059CPA*D $-$ 6.50

$$(n = 70, R^2 = 0.55, P <0.001)$$

Emulsifying-Stability Index ESI = 0.38D $-$ 0.012ANS*D + 0.00015CPA*SH $-$ 0.28

$$(n = 56, R^2 = 0.43, P <0.001)$$

Note: Quadratic equations were computed by backward stepwise multiple regression; D = dispersibility; ANS = hydrophobicity determined using ANS; CPA = hydrophobicity determined using CPA; SH = sulfhydryl content. Samples included unheated and heated (35 to 75°C) extracts as well as extracts analyzed after frozen storage at -10°C.

Adapted from Li-Chan, E., Nakai, S., and Wood, D. F.[49] *J. Food Sci.*, 50, 1034, 1985.

QSAR approach and multivariate data analyses to predict muscle protein functionality from physicochemical properties, the significant findings are reported in detail here, in the hope of stimulating other researchers to discover the greater potential application of the QSAR approach for understanding and controlling the functional quality parameters of food proteins, through basic knowledge of their structure and physicochemical properties.

Insight into the muscle protein structure-function relationship on a quantitative level was sought by multivariate analyses of data from a variety of hand-deboned and mechanically deboned samples, including beef, chicken, pork, and fish. Functional properties under evaluation included properties of the meat mince (water- and brine-holding capacity, gel strength, cookloss) as well as of the salt-extracted proteins (gel strength, emulsifying, and fat-binding capacities). Physicochemical properties included variables related to the mince (protein, salt-extractable protein, fat, moisture, and pH), and variables describing salt-extractable protein properties (solubility, dispersibility, hydrophobicity, and sulfhydryl content). Table 16 summarizes the types of samples and properties which were included in the analyses.

To obtain an overview of the general relationship between functional and physicochemical properties of the meat samples, stepwise multiple regression analyses using quadratic models were carried out. Each functional property was considered in turn to be the dependent variable, while the physicochemical properties were entered as potential independent predictor variables. These analyses yielded empirical equations which could be readily applied to predict functional properties of unknown samples based on measurement of their physicochemical properties (Table 17). Inclusion of various quadratic terms of the properties in these equations supported earlier suggestions[48,49] that the relationships between functionality and physicochemical properties of proteins are rarely linear, and an optimum balance of various properties is required for good functionality.

Thermally induced properties included Gel-M, Gel-E, cookloss and water-binding properties describing retained moisture after cooking samples in water or brine (RM$_w$ and RM$_b$), which were measured after 40°C/80°C or 80°C thermal treatment. For these thermally induced properties, it was noted that significant predictor variables included not only the physico-

Table 16
TYPES OF SAMPLES, FUNCTIONAL AND
PHYSICOCHEMICAL PROPERTIES OF MEAT
SAMPLES FOR MULTIVARIATE DATA ANALYSES[a]

Samples (fresh and frozen)
    Hand-deboned   P = pork (lean ham trim)
                                   C = chicken (broiler breast)
                                   B = beef (top round)
                                   F/f = fish (ling cod ± cryoprotectants)
    Mechanically deboned   M = pork
                                   m = chicken (backs and necks)
Physicochemical properties
    Mince   Moisture, %
                                   Fat, %
                                   Protein, %
                                   Salt-extractable protein, % (SEP)
                                   pH
    Salt extract   Solubility, % (27,000 × g, 30 min)
                                   Dispersibility, % (1100 × g, 10 min)
                                   Hydrophobicity (ANS)
                                   Hydrophobicity (CPA)
                                   Sulfhydryl group, $\mu M/g$ (SH)
Functional properties
    Mince   Gel strength, N (Gel-M)
                                   Cookloss, %
                                   Water-binding capacity, g/g sample
                                     Uncooked AM_{water}, AM_{brine})
                                     Cooked (RM_{water}, RM_{brine})
    Salt extract   Gel strength, N (Gel-E)
                                   Emulsifying capacity, g oil/30 mg (EC)
                                   Fat-binding capacity, mℓ oil/g (FBC)

[a]   Details of materials and methods are in Li-Chan et al.[53]

chemical properties of the heated samples, but also those of the samples prior to heating (designated as "property-u" for unheated) as well as the ratio of heated:unheated property (designated as "property-ratio").

Although the relationships described by multiple regression analyses offer a simple approach easily adapted by industry for prediction of functionality, the reliability or accuracy of this approach may be reduced if the properties used as independent predictor variables in fact are correlated to each other. In such cases, multivariate analyses of data using techniques such as factor or principal component analysis may be used to circumvent the problem of multicollinearity of the variables.

Information on the correlation or extent of similarity between the various physical/chemical and functional properties was gathered by cluster analysis of these variables from data of 230 samples, using the amalgamation procedure based on single linkage. The measures of similarity were derived by recoding the absolute values of the correlation matrix of the variables (Table 18). The dendogram shows that the variables fell broadly into two clusters (Figure 14, I and II). In the first broad cluster (I), solubility and dispersibility formed a tight cluster of high similarity (94%) which was further clustered with emulsifying capacity (86% similarity). This cluster of three highly correlated variables was in turn clustered to variables describing gel strength (Gel-M, Gel-E), which were next combined with a cluster describing protein hydrophobicity properties (ANS, CPA). These seven variables formed a tight cluster of high similarity (77%) which was related to the hydrophobic and hydrophilic properties of the meat proteins and their gel-forming ability. This cluster was linked with

Table 17
RELATIONSHIPS BETWEEN FUNCTIONAL PROPERTIES AND PHYSICOCHEMICAL PROPERTIES OF MEAT SAMPLES

Thermally induced properties

(1) Gel-M $= -0.029$ ANS $- 0.38$ fat $- 3.15$ mince-pH $+ 0.17$ dispersibility-u $+ 0.000024$ ANS^2 $- 2.70$ SEP^2 $- 0.0018$ moisture^2 $+ 0.0085$ solubility $\cdot$ SH $+ 0.44$ SEP·protein $+ 27.24$ (n $= 114$, $R^2 = 0.8344$, S.E. $= 1.9609$, $F_{(9,104)} = 58.21$)

(2) Cookloss $= 0.010$ ANS $- 77.3$ SEP $- 1.61$ moisture $+ 0.80$ fat $+ 30.08$ SEP^2 $- 0.062$ fat^2 $- 0.35$ mince pH^2 $- 0.00065$ solubility·CPA $+ 0.52$ CPA·SH $+ 0.79$ SEP·solubility $+ 173.9$ (n $= 114$, $R^2 = 0.8952$, S.E. $= 3.0525$, $F_{(10,103)} = 87.97$)

(3a)[a] Gel-E $= -0.051$ SEP $+ 0.000024$ CPA-u $+ 0.000012$ dispersibility^2 $- 0.083$ dispersibility-ratio $+ 0.047$ SH-ratio $- 0.00000038$ solubility·CPA $+ 0.000000011$ ANS·CPA $+ 0.0033$ SEP·protein $- 0.027$ (n $= 114$, $R^2 = 0.7634$, S.E. $= 0.0087$, $F_{(8,105)} = 42.34$)

(3b)[a] Gel-E $= -0.00016$ ANS-u $+ 0.000014$ CPA-u $+ 0.00070$ SH-u $- 0.019$ (n $= 114$, $R^2 = 0.5596$, S.E. $= 0.0116$, $F_{(3,110)} = 46.59$)

(4) RM$_w$ $= 0.015$ dispersibility $- 0.013$ dispersibility-u $- 0.00019$ CPA-u $- 0.00012$ dispersibility^2 $+ 0.088$ SEP^2 $+ 1.50$ (n $= 58$, $R^2 = 0.6931$, S.E. $= 0.0948$, $F_{(5,52)} = 23.49$)

(5) RM$_b$ $= 0.57$ SEP $+ 0.0086$ solubility-u $- 0.018$ dispersibility-u $+ 0.00015$ fat^2 $- 0.00000065$ CPA·SH $- 0.0074$ SEP·sol $- 0.0087$ SEP·protein $+ 1.13$ (n $= 58$, $R^2 = 0.7287$, S.E. $= 0.1070$, $F_{(7,50)} = 19.19$)

Unheated properties

(6) AM$_w$ $= 0.014$ ANS $- 0.024$ protein $+ 0.39$ mince-pH $+ 0.00012$ moisture^2 $- 0.0000096$ CPA·SH $- 2.06$ (n $= 58$, $R^2 = 0.8894$, S.E. $= 0.1152$, $F_{(5,52)} = 83.67$)

(7) AM$_b$ $= 0.002064$ CPA $- 0.0000017$ CPA^2 $- 0.26$ (n $= 58$, $R^2 = 0.4390$, S.E. $= 0.0949$, $F_{(2,55)} = 21.52$)

General properties

(8a)[a] EC $= 0.091$ dispersibility $- 0.024$ CPA $- 0.000046$ ANS^2 $+ 0.0015$ moisture^2 $+ 0.00028$ sol·CPA $+ 0.00040$ ANS·CPA $+ 33.26$ (n $= 230$, $R^2 = 0.7740$, S.E. $= 6.3429$, $F_{(6,223)} = 127.28$)

(8b)[a] EC $= 0.30$ dispersibility $+ 0.00011$ ANS·SH $+ 21.51$ (n $= 230$, $R^2 = 0.7411$, S.E. $= 6.7281$, $F_{(2,227)} = 324.96$)

(9a)[a] FBC $= -16.2$ SEP $- 0.00029$ solubility^2 $+ 0.00034$ SH^2 $+ 7.6$ SEP^2 $+ 0.073$ SEP·solubility $+ 8.22$ (n $= 230$, $R^2 = 0.5560$, S.E. $= 2.2981$, $F_{(5,224)} = 56.09$)

(9b)[a] FBC $= -0.00093$ solubility^2 $+ 0.0017$ solubility·SH $+ 2.19$ (n $= 230$, $R^2 = 0.3737$, S.E. $= 2.7111$, $F_{(2,227)} = 67.73$)

Note: F-values for all the regression equations were significant at $P < 0.001$.

[a] For Gel-E, EC, and FBC, the functional properties were measured on the salt extracts of proteins after adjustment to specified conditions or protein concentration and pH. Regression equations for these properties were obtained using both mince and extract properties (Equations 3a, 8a, and 9a) or using extract properties only (Equations 3b, 8b, and 9b) as the potential independent variables. (From Li-Chan, E., Nakai, S., and Wood, D. F., *J. Food Sci.*, 1986, 52, 31, 1987. With permission.)

cookloss and fat-binding capacity properties with moderate (50%) similarity, which supports the hypothesis that these two properties may be related to the formation of a protein gel matrix.

The second broad cluster (II) identified from the dendogram appears to be generally related to the compositional properties of the mince and also sulfhydryl content of the proteins. Within this broad cluster were a cluster of three variables composed of total and salt-extractable protein and mince-pH, and a tight cluster of fat and moisture contents which were highly negatively correlated (-0.91) to each other.

The two broad clusters of variables in the dendogram are only correlated to each other with moderate similarity (38%), which may suggest that compositional properties of meat mince are poor predictors of functional properties such as gel strength, cookloss, and emulsifying- and fat-binding capacities.

Table 18

CORRELATION MATRIX OF FUNCTIONAL AND PHYSICOCHEMICAL PROPERTIES OF MUSCLE SAMPLES[52]

	Gel-M	Cook-loss	Gel-E	EC	FBC	Solub.	Dis-pers.	ANS	CPA	SH	SEP	Mois-ture	Fat	Protein	Mince-pH
Gel-M	1.000														
Cookloss	0.230	1.000													
Gel-E	0.752	0.501	1.000												
EC	−0.671	−0.486	−0.654	1.000											
FBC	−0.461	−0.275	−0.435	0.455	1.000										
Solubility	−0.784	−0.489	−0.736	0.827	0.528	1.000									
Dispers.	−0.814	−0.537	−0.794	0.856	0.513	0.943	1.000								
ANS	0.754	0.353	0.747	−0.601	−0.432	−0.754	−0.766	1.000							
CPA	0.654	0.444	0.703	−0.656	−0.487	−0.721	−0.770	0.816	1.000						
SH	−0.150	−0.077	−0.002	0.279	0.377	0.310	0.242	−0.032	−0.132	1.000					
SEP	0.304	−0.237	0.264	−0.079	0.139	−0.061	−0.062	0.097	−0.067	0.384	1.000				
Moisture	0.171	−0.356	−0.045	0.132	0.241	−0.040	0.040	0.216	−0.162	0.197	0.271	1.000			
Fat	−0.263	0.279	−0.063	−0.094	−0.241	0.054	−0.011	−0.246	0.159	−0.374	−0.491	−0.914	1.000		
Protein	0.294	−0.034	0.353	−0.122	0.154	−0.034	−0.120	0.050	−0.036	0.370	0.703	0.016	−0.247	1.000	
Mince-pH	−0.324	0.087	−0.245	0.093	−0.191	0.046	0.100	−0.146	0.035	−0.449	−0.694	−0.362	0.593	−0.757	1.000

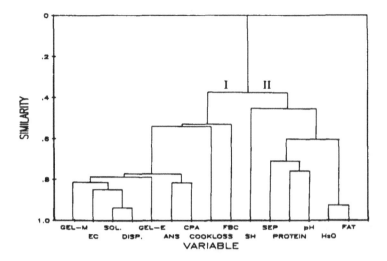

FIGURE 14. Dendogram obtained from cluster analysis of variables describing physical, chemical, and functional properties of meat samples. (Adapted from Li-Chan, E., Nakai, S., and Wood, D. F., *J. Food Sci.*, 52, 31, 1987. With permission.)

Factor analysis was also used for exploratory data analysis to study the interrelationships between the variables or properties and to identify the minimum number of factors or principal components which could account for the variation in the data set. Since properties with high loadings on the same factor tend to be highly correlated with each other, while properties with dissimilar loading patterns tend to be less highly correlated, the problem of multicollinearity is overcome by considering these factors rather than individual properties.

Three factors identified by principal component analysis accounted for over 76% of variance in 230 heated and unheated samples. From the sorted rotated factor loadings pattern (Table 19), it may be seen that factor 1 was related to hydrophobic and hydrophilic protein properties and gel strength, being comprised of positive values of ANS, CPA, Gel-M, and Gel-E, and negative values of dispersibility, solubility, and EC. Factor 2 was described by positive values of total and salt-extractable protein contents and a negative term in mince-pH, while factor 3 was described by a positive term in moisture content and a negative term in fat content.

Identification of the location of individual samples on plots of factor scores of samples (Figures 15, A and B) indicated that unheated or 40°C-heated samples were located on the negative scale of the factor 1 axis while 80°C- and 40/80°C-heated samples were on the positive scale of the factor 1 axis. Although they exhibited considerable variability in their factor 1 scores, hand-deboned chicken samples (C) generally showed the largest difference between factor 1 scores of unheated and heated samples, reflecting the high gel strength, large increase in hydrophobicity, and decrease in solubility upon heating of these samples.

Three groups were distinguished from plots of the factor scores of the samples (Figures 15A, B, and C). Group 1 included hand-deboned chicken, beef, and pork (C, B, and P) samples. It was characterized by a large range in factor 1 scores between unheated (negative score) and heated (positive score) samples, reflecting their high gel strength, increase in hydrophobicity, and decrease in solubility after heating. Group 1 samples had positive factor 2 scores (high protein and salt-extractable protein, and low pH) and intermediate factor 3 scores (moderate moisture and fat contents). Group 2 included mechanically deboned chicken and pork (m and M) samples and was characterized by highly negative factor 3 scores (low moisture and high fat contents) and intermediate values for the range of factor 1 scores, and intermediate factor 2 scores. Group 3 included the fish samples (F and f) and was charac-

Table 19
SORTED ROTATED FACTOR
LOADINGS (PATTERN) FROM FACTOR
ANALYSIS OF MEAT DATA
EXCLUDING WBC

	Factor 1	Factor 2	Factor 3
Dispersibility	−0.959	0.0	0.0
Solubility	−0.939	0.0	0.0
ANS	0.864	0.0	0.255
Gel-M	0.862	0.0	0.0
Gel-E	0.853	0.288	0.0
EC	−0.848	0.0	0.0
CPA	0.844	0.0	0.0
FBC	−0.604	0.293	0.0
Cookloss	0.529	0.0	−0.457
Protein	0.0	0.916	0.0
Mince-pH	0.0	−0.852	0.285
S.E.P.	0.0	0.821	0.0
SH	0.0	0.617	0.0
Moisture	0.0	0.0	0.966
Fat	0.0	−0.382	−0.891

Note: The above factor-loading matrix has been rearranged
so that the columns appear in decreasing order of
variance explained by factors. The rows have been
rearranged so that for each successive factor loadings
greater than 0.5000 appear first. Loadings less than
0.2500 have been replaced by zero.

Adapted from Li-Chan, E., Nakai, S., and Wood, D. F.,
J. Food Sci., 1986. 52, 31, 1987.

terized by a small range of factor 1 scores (reflecting minor changes by heating), negative
factor 2 scores (low protein content, high pH), and highly positive factor 3 scores (high
moisture and low fat contents).

When factor analysis was performed on the data subset of only unheated samples, including
absorbed moisture variables (AM_w and AM_b), four factors were identified which accounted
for over 77% of the variance in the data. The sorted rotated factor loadings pattern (Table
20) shows that factor 1 had high loadings on fat, CPA, and pH variables and negative
loadings on moisture, SH and SEP. Factor 2 had high loadings on AM_w and ANS hydrophobicity and negative loading on protein. Factor 3 showed heavy loadings on solubility,
dispersibility, and EC, while factor 4 was described mainly by AM_b.

The plots of the factor scores of these samples showed that the cod fish samples (F and
f) were easily distinguished by their high factor 2 scores reflecting their high AM_w and ANS
hydrophobicity and low protein (Figure 16A). Mechanically-deboned (m and M) and hand-
deboned (C, B, and P) samples had similar scores for factor 2 but could be distinguished
by their factor 1 scores (Figure 16A), reflecting the higher fat, CPA, and pH, and lower
moisture and salt-extractable proteins in the mechanically deboned samples. The plot in
Figure 16B shows that factor 3 could not be used to distinguish between species or type of
deboning, due to the large variability in solubility, dispersibility, and EC common to all
these types of samples.

To investigate the possibility of discriminating between samples of different functionality
based on their physicochemical properties, stepwise discriminant analyses of the data were
performed. For each functional property, samples were assigned into one of three groups

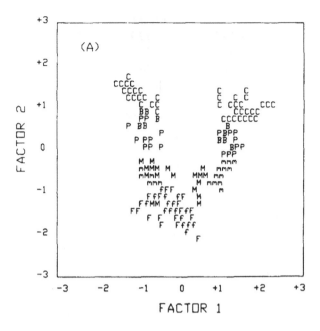

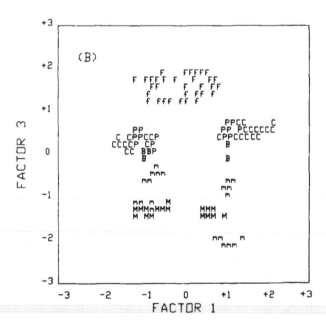

FIGURE 15.   Plots of factor scores of unheated and heated meat samples obtained by principal component analysis of all variables excluding water-binding capacity: (A) factor 2 vs. factor 1; (B) factor 3 vs. factor 1; (C) factor 3 vs. factor 2. Sample abbreviations: C, hand-deboned chicken; B, hand-deboned beef; P, hand-deboned pork; M, mechanically deboned chicken; F, cod fish; f, cod fish in presence of cryoprotectants. Overlapping factor scores are not shown. (From Li-Chan, E., Nakai, S., and Wood, D. F., *J. Food Sci.*, 52, 31, 1987. With permission.)

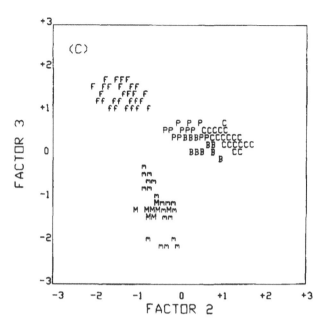

Table 20

**SORTED ROTATED FACTOR LOADINGS
(PATTERN) FROM FACTOR ANALYSIS OF
MEAT DATA INCLUDING AM$_w$ AND AM$_b$**

	Factor 1	Factor 2	Factor 3	Factor 4
Fat	0.951	0.0	0.0	0.0
Moisture	−0.832	0.453	0.0	0.0
CPA	0.827	0.0	0.0	−0.277
pHab	0.826	0.460	0.0	0.0
pHaw	0.803	0.469	0.0	0.0
Mince-pH	0.747	0.566	0.0	0.0
SH	−0.739	0.0	0.0	−0.265
S.E.P.	−0.605	−0.529	0.0	0.0
AM$_w$	0.0	0.926	0.0	0.306
ANS	0.0	0.898	0.0	0.0
Protein	−0.447	−0.818	0.0	0.0
Solubility	0.0	0.0	0.766	0.0
Dispersibility	0.0	0.0	0.758	0.0
EC	0.0	0.0	0.574	0.0
AM$_b$	0.0	0.0	0.0	0.955
FBC	−0.414	−0.351	0.0	0.380

Note: The above factor-loading matrix has been rearranged so that the
columns appear in decreasing order of variance explained by
factors. The rows have been rearranged so that for each successive
factor loadings greater than 0.5000 appear first. Loadings less
than 0.2500 have been replaced by zero.

Adapted from Li-Chan, E., Nakai, S., and Wood, D. F., *J. Food Sci.*,
1986. 52, 31, 1987. With permission.

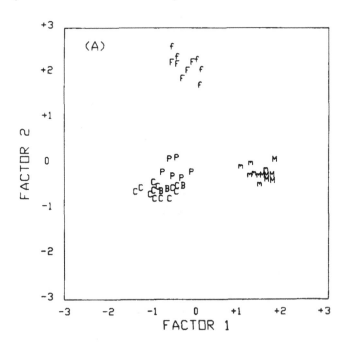

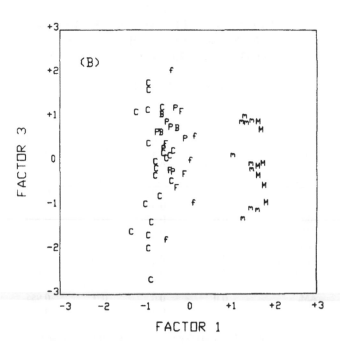

FIGURE 16.   Plots of factor scores for unheated meat samples obtained by principal component analysis: (A) factor 2 vs. factor 1; (B) factor 3 vs. factor 1. Sample abbreviations are as indicated in the caption to Figure 15. (From Li-Chan, E., Nakai, S., and Wood, D. F., *J. Food Sci.*, 52, 31, 1987. With permission.)

representing low (A), medium (B), and high (C) functionality. Stepwise discriminant analysis was used to determine the subset of physicochemical properties necessary to describe the classification functions which could best discriminate between the groups. The coefficients for the variables entered into the classification functions for classifying the three groups for each functional property are shown in Table 21. The coefficients of the two canonical variables used for visual discrimination of the groups in a two-dimensional space are also listed in Table 21, the first canonical variable being the linear combination of predictor variables best discriminating among groups along the X-axis, and the second canonical variable discriminating among groups along the Y-axis. The canonical plots of the individual samples as well as the means of the three groups are shown for each functional property (Figure 17A through H). The percentages of correct classification of samples into the three groups according to the classification functions for each functional property ranged from 62.5% for FBC to 98.3% for AM_w classification.

These results of stepwise discriminant analyses demonstrated that samples could be classified into groups of low, medium, and high functionality, with generally good accuracy of classification. The classification functions shown in Table 21 could be used to predict the category of functional property of unknown samples, based on their properties as represented by the selected variables in the classification functions. As an alternative, the values for the first two canonical variables could be calculated for unknown samples using the coefficients for the selected variables shown in Table 21, and the location of the unknowns on the canonical plots (Figure 17A through H) could then be used to compare the unknown samples' predicted functionality to that of known samples.

The capability for obtaining quantitative information on muscle protein structure-function relationships and discrimination of functionality by multivariate analyses has thus been demonstrated. The great challenge is for wider acceptance of this approach by both basic and applied researchers and transfer of the knowledge to the meat processing industry. By checking the quality of actual products which have been formulated after incorporation of predicted functionality as constraints in linear programming, the prediction equations and classification functions should be improved and refined by comparison of the predicted and observed functionality. Using the QSAR approach in cooperation with refinement by practical experience, it is anticipated that linear programming for least cost product formulation can be significantly improved by incorporation of predicted ingredient functionality as constraints to maintain constant product quality.

Table 21

CLASSIFICATION FUNCTIONS FOR LOW, MEDIUM, AND HIGH GROUPS OF MEAT FUNCTIONAL PROPERTIES AND COEFFICIENTS FOR FIRST TWO CANONICAL VARIABLES

Grouping variable	Variable in classification function	Coefficient for classification functions			Coefficient for	
		Group 1 (low)	Group 2 (medium)	Group 3 (high)	Canonical variable 1	Canonical variable 2
Gel-M	Solubility	−1.881	−2.558	−1.373	−0.0898	−0.2961
	SEP	111.6	159.4	151.8	−7.9059	4.6444
	ANS-u	−0.064	0.013	−0.026	−0.0096	0.0137
	CPA-u	0.117	0.094	0.092	0.0055	−0.0012
	Solubility^2	0.028	0.032	0.021	0.0013	0.0024
	SEP^2	−57.94	−77.28	−69.03	2.6867	−3.1357
	Moisture^2	0.052	0.050	0.048	0.0009	0.0003
	Fat^2	0.181	0.162	0.152	0.0063	0.0008
	Dispersibility-ratio	−25.34	−3.17	−36.82	1.8823	8.6234
	Constant	−204.9	−203.2	−189.5	—	—
Cookloss	Solubility	−0.045	0.505	0.531	−0.0578	−0.0420
	Dispersibility	0.704	−0.732	−1.251	0.185	0.0344
	SEP	951.8	856.4	777.9	15.59	−5.09
	Moisture	21.04	21.48	19.13	0.131	−0.428
	Mince-pH	248.1	225.4	224.2	2.400	1.725
	ANS-u	0.604	0.454	0.391	0.0200	0.0021
	CPA-u	−0.210	−0.177	−0.152	−0.0052	0.0014
	Dispersibility^2	−0.027	−0.019	−0.013	−0.0013	0.0003
	SEP^2	−405.8	−364.5	−332.1	−6.63	1.94
	Fat^2	0.329	0.353	0.292	0.0022	−0.0122
	Constant	−1879.2	−1689.2	−1478.5	—	—
Gel-E	Dispersibility	1.918	1.071	0.931	0.2264	−0.0384
	SH	0.441	0.607	0.653	−0.0473	−0.0134
	ANS-u	0.041	−0.0034	−0.0053	0.0109	−0.0080
	Dispersibility^2	−0.017	−0.010	−0.0086	−0.0020	0.0002
	ANS-ratio	0.775	0.0934	0.370	0.1197	−0.4647
	Constant	−38.22	−27.76	−31.14	—	—
RM$_w$	Dispersibility-u	6.428	6.140	5.953	0.1277	0.0520
	Dispersibility^2	0.015	0.008	0.009	0.0021	0.0028
	SEP^2	−7.227	−2.949	−1.979	−1.6065	−13.2523
	Dispersibility-ratio	−147.5	−82.32	−71.73	−23.7933	−214.5
	Constant	−308.2	−291.1	−281.6	—	—
RM$_b$	CPA	0.028	0.021	0.020	0.0005	−0.0124
	SEP	−28.49	−13.52	−9.082	−1.0685	8.2003
	Solubility-u	−0.479	0.167	0.034	−0.0256	−0.6550
	Dispersibility-u	7.726	7.203	6.678	0.0486	2.5350
	Dispersibility^2	0.008	0.006	0.005	0.0002	−0.0018
	Constant	−374.9	−351.9	−323.1		
AM$_w$	ANS	0.620	0.639	0.813	−0.0171	−0.0002
	SEP	−8.585	3.693	−20.32	1.1501	−2.2799
	Moisture	20.36	22.19	23.55	−0.2698	−0.2627
	Protein	−2.564	−4.290	−6.190	0.3100	0.2383
	Mince-pH	321.2	317.8	376.3	−4.9557	1.4414
	CPA^2	−0.00019	−0.00021	−0.00025	0.00001	0.00000
	Fat^2	0.248	0.368	0.262	−0.00032	−0.02030
	Constant	−1683.1	−1801.1	−2231.5		
EC	Solubility	0.129	0.152	0.225	−0.0257	−0.0008
	ANS	0.019	0.036	0.040	−0.0024	−0.0112
	Dispersibility^2	0.00007	0.00038	0.00099	−0.0002	−0.0001
	CPA^2	0.00000	0.00000	−0.00000	0.0000	0.0000
	Constant	−7.235	−9.975	−17.30		

Table 21 (continued)
CLASSIFICATION FUNCTIONS FOR LOW, MEDIUM, AND HIGH GROUPS OF MEAT FUNCTIONAL PROPERTIES AND COEFFICIENTS FOR FIRST TWO CANONICAL VARIABLES

Grouping variable	Variable in classification function	Coefficient for classification functions			Coefficient for	
		Group 1 (low)	Group 2 (medium)	Group 3 (high)	Canonical variable 1	Canonical variable 2
FBC	Solubility	0.142	0.173	0.162	−0.0164	0.0361
	CPA	0.017	0.016	0.014	0.0013	0.0021
	SH²	0.0012	0.0014	0.0018	−0.00024	−0.00043
	Constant	−15.45	−17.08	−18.03		

Adapted from Li-Chan, E., Nakai, S., and Wood, D. F., *J. Food Sci.*, 1986. 52, 31, 1987.

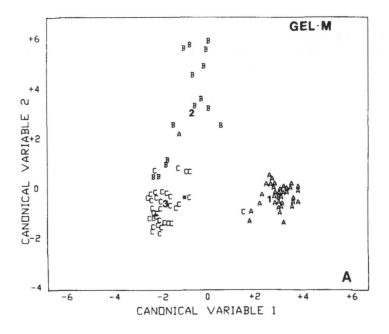

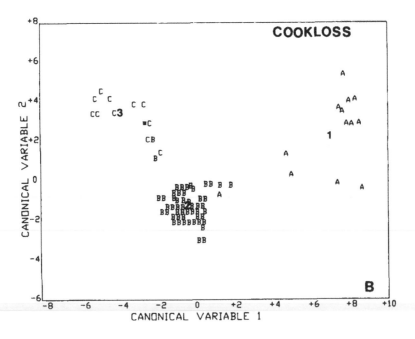

FIGURE 17.   Canonical plots of meat samples classified into 3 groups of low (A), medium (B) and high (C) functionality and their corresponding group means 1, 2, and 3. Overlap of different groups is indicated by *. Canonical variables were computed from variables selected by stepwise discriminant analyses for each functional property: (A) Gel-M, (B) cookloss, (C) Gel-E, (D) RM_w, (E) RM_b, (F) AM_w (G) EC, and (H) FBC. (From Li-Chan, E., Nakai, S., and Wood, D. F., *J. Food Sci.*, 52, 31, 1987.With permission.)

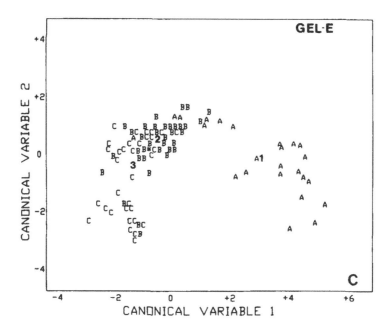

FIGURE 17C

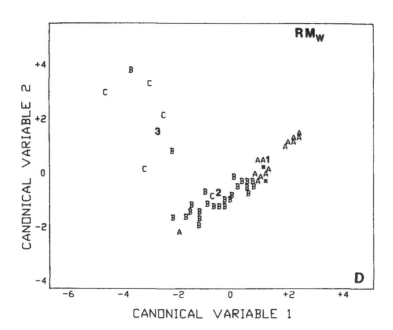

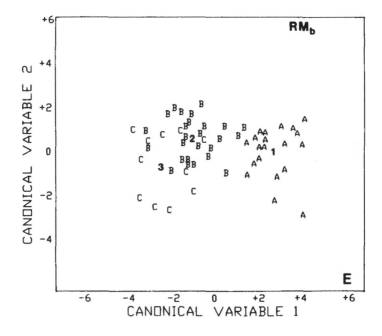

FIGURE 17E

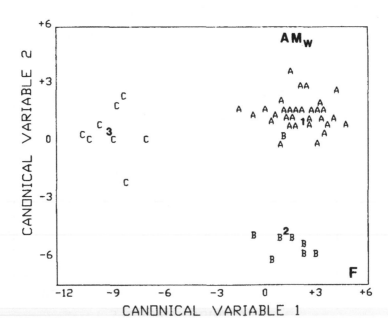

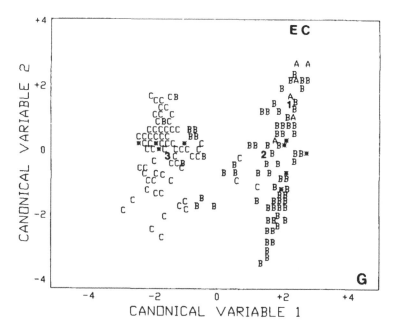

FIGURE 17G

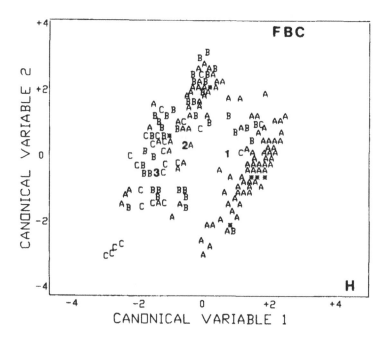

REFERENCES

1. **Hultin, H. O.**, Characteristics of muscle tissue, in *Principles of Food Science, Part I. Food Chemistry*, Fennema, O. R., Ed., Marcel Dekker, New York, 1976, 577.
2. **Obinata, T., Masaki, T., and Takano, H.**, Immunochemical comparison of myosin light chains from chicken fast white, slow red and cardiac muscle, *J. Biochem.*, 86, 131, 1979.
3. **Samejima, K., Hara, S., Yamamoto, K., Ashgar, A., and Yasui, T.**, Physicochemical properties and heat-induced gelling of cardiac myosin in model system, *Agric. Biol. Chem.*, 49, 2975, 1985.
4. **Ashgar, A., Samejima, K., and Yasui, T.**, Functionality of muscle proteins in gelation mechanisms of structured meat products, *CRC Crit. Rev. Food Sci. Nutr.*, 22, 27, 1986.
5. **Ashgar, A. and Yeates, N. T. M.**, The mechanism for the promotion of tenderness in meat during the post-mortem process: a review, *CRC Crit. Rev. Food Sci. Nutr.*, 10, 115, 1978.
6. **Lawrie, R. A.**, in *Meat Science*, 4th ed., Lawrie, R. A., Ed., Pergamon Press, New York, 1985, 4.
7. **Porteous, J. D.**, Some physico-chemical "constants" of various meats for optimum sausage formulation, *Can. Inst. Food Sci. Technol. J.*, 12, 145, 1979.
8. **Parks, L. L., Carpenter, J. A., Rao, V. N. M., and Reagan, J. O.**, Prediction of bind value constants of sausage ingredients from protein or moisture content, *J. Food Sci.*, 50, 1564, 1985.
9. **Saffle, R. L.**, Meat emulsions, *Adv. Food Res.*, 16, 105, 1968.
10. **Lee, C. M.**, Mechanisms of fat dispersion in comminuted muscle protein matrices, in *Engineering and Food*, Vol. 1, McKenna, B. M., Ed., Elsevier, New York, 1984, 403.
11. **Comer, F. W.**, Functionality of fillers in comminuted meat products, *Can. Inst. Food Sci. Technol. J.*, 12, 157, 1979.
12. **Comer, F. W. and Dempster, S.**, Functionality of fillers and meat ingredients in comminuted meat products, *Can. Inst. Food Sci. Technol. J.*, 14, 295, 1981.
13. **Trinick, J. A. and Cooper, J.**, AMP deaminase: its binding and location within rabbit psoas myofibrils, *J. Muscl. Res. Cell. Mobil.*, 3, 486, 1982.
14. **Yates, L. D. and Greaser, M. L.**, Quantitative determination of myosin and actin in rabbit skeletal muscle, *J. Mol. Biol.*, 168, 123, 1983.
15. **McLachlan, A. D. and Stuart, M.**, Tropomyosin coiled-coil interactions: evidence for an unstaggered structure, *J. Mol. Biol.*, 98, 293, 1975.
16. **Goll, D. E., Robson, R. M., and Stromer, M. H.**, Muscle Proteins, in *Food Proteins*, Whitaker, J. R. and Tannenbaum, S. R., Eds., AVI Publishing, Westport, Conn., 1977, 121.
17. **Tonomura, Y.**, *Muscle Proteins, Muscle Contraction and Cation Transport*, University of Tokyo Press, Tokyo. Translated by T. Takeshita, University Park Press, Baltimore, 1973.
18. **Bodwell, C. E. and McClain, P. E.**, Chemistry of Animal Tissues — Proteins, in *The Science of Meat and Meat Products*, Price, J. F. and Schweigert, B. S., Eds., W. H. Freeman, San Francisco, 1971, 78.
19. **Chung, C.-S., Richards, E. G., and Olcott, H. S.**, Purification and properties of tuna myosin, *Biochemistry*, 6, 3154, 1967.
20. **Sarkar, S., Sreter, F. A., and Gergely, J.**, Light chains of myosins from white, red and cardiac muscles, *Proc. Natl. Acad. Sci. U.S.A.*, 68, 946, 1971.
21. **Lowey, S. and Risby, D.**, Light chains from fast and slow muscle myosins, *Nature*, 234, 81, 1971.
22. **Masaki, T.**, Immunochemical comparison of myosins from chicken cardiac, fast white, slow red, and smooth muscle, *J. Biochem.*, 76, 441, 1974.
23. **Angel, S. and Weinberg, Z. G.**, Gelation property of salt soluble protein of turkey muscle as related to pH, *J. Food Technol.*, 16, 549, 1981.
24. **Ashgar, A., Morita, J. I., Samejima, K., and Yasui, T.**, Biochemical and functional characteristics of myosin from red and white muscles of chicken as influenced by nutritional stress, *Agric. Biol. Chem.*, 48, 2217, 1984.
25. **Ebashi, S. and Nonomura, Y.**, Proteins of the myofibril, in *The Structure and Function of Muscle*, Vol. 3, Bourne, G. H., Ed., Academic Press, New York, 1973, 285.
26. **Weeds, A.**, Actin-binding proteins — regulators of cell architecture and motility, *Nature*, 296, 811, 1982.
27. **Hofmann, K. and Hamm, R.**, Sulfhydryl and disulfide groups in meats, *Adv. Food Res.*, 24, 2, 1978.
28. **Lowey, S., Slayter, H. S., Weeds, A. G., and Baker, H.**, Substructure of the myosin molecule. I. Subfragments of myosin by enzymic degradation, *J. Mol. Biol.*, 42, 1, 1969.
29. **Tonomura, Y., Kazuko, S., and Imamura, K.**, The optical-rotatory dispersion of myosin A. II. Effect of dioxane and *p*-chloromercuribenzoate, *Biochim. Biophys. Acta*, 69, 296, 1963.
30. **Duke, J. A., McKay, R., and Botts, J.**, Conformational change accompanying modification of myosin ATPase, *Biochim. Biophys. Acta*, 126, 600, 1966.
31. **Lim, S. T. and Botts, J.**, Temperature and aging effects on the fluorescence intensity of myosin-ANS complex, *Arch. Biochem. Biophys.*, 122, 153, 1967.

32. **Cheung, H. C. and Morales, M. F.**, Studies of myosin conformation by fluorescent techniques, *Biochemistry*, 8, 2177, 1969.

33. **Förster, T.**, Transfer mechanisms of electronic excitation, *Discussions Faraday Soc.*, 27, 7, 1959.

34. **Cheung, H. C.**, Conformation of myosin — effects of substrates and modifiers, *Biochim. Biophys. Acta*, 194, 478, 1969.

35. **Borejdo, J.**, Mapping of hydrophobic sites on the surface of myosin and its fragments, *Biochemistry*, 22, 1182, 1983.

36. **Pinaev, G., Tartakovsky, A., Shanbhag, V. P., Johansson, G., and Backman, L.**, Hydrophobic surface properties of myosin in solution as studied by partition in aqueous two-phase systems: effects of ionic strength, pH and temperature, *Mol. Cell. Biochem.*, 48, 65, 1982.

37. **Burtnick, L. D. and Chan, K. W.**, Fluorescence of actin-bound hydrophobic molecules, *Can. J. Biochem. Cell. Biol.*, 61, 981, 1983.

38. **Asakura, S., Kasai, M., and Oosawa, F.**, The effect of temperature on the equilibrium state of actin solutions, *J. Polym. Sci.*, 44, 35, 1960.

39. **Gordon, D. J., Yang, Y. Z., and Korn, E. D.**, Polymerization of *Acanthamoeba* actin. Kinetics, thermodynamics, and co-polymerization with muscle actin, *J. Biol. Chem.*, 251, 7474, 1976.

40. **Niwa, E.**, Role of hydrophobic bonding in gelation of fish flesh paste, *Bull. Jpn. Soc. Sci. Fish.*, 41, 907, 1975.

41. **Niwa, E., Koshiba, K., Matsuzaki, M., Nakayama, T., and Hamada, I.**, Species-specificities of myosin heavy chain in setting and returning, *Bull. Jpn. Soc. Sci. Fish.*, 46, 1497, 1980.

42. **Niwa, E., Suzuki, R., and Hamada, I.**, Fluorometry of the setting of fish flesh sol supplement, *Bull. Jpn. Soc. Sci. Fish.*, 47, 1389, 1981.

43. **Niwa, E., Sato, K., Suzuki, R., Nakayama, T., and Hamada, I.**, Fluorometric study of setting properties of fish flesh sol, *Bull. Jpn. Soc. Sci. Fish.*, 47, 817, 1981.

44. **Niwa, E., Matsubara, Y., and Hamada, I.**, Hydrogen and other polar bondings in fish flesh gel and setting gel, *Bull. Jpn. Soc. Sci. Fish.*, 48, 667, 1982.

45. **Niwa, E., Nakayama, T., and Hamada, I.**, Effect of arylation for setting of muscle proteins, *Agric. Biol. Chem.*, 45, 341, 1981.

46. **Niwa, E., Nakayama, T., and Hamada, I.**, Arylsulfonyl chloride induced setting of dolphinfish flesh sol, *Bull. Jpn. Soc. Sci. Fish.*, 47, 179, 1981.

47. **Niwa, E., Suzuki, R., Sato, K., Nakayama, T., and Hamada, I.**, Setting of flesh sol induced by ethylsulfonation, *Bull. Jpn. Soc. Sci. Fish.*, 47, 915, 1981.

48. **Li-Chan, E., Nakai, S., and Wood, D. F.**, Hydrophobicity and solubility of meat proteins and their relationship to emulsifying properties, *J. Food Sci.*, 49, 345, 1984.

49. **Li-Chan, E., Nakai, S., and Wood, D. F.**, Relationship between functional (fat binding, emulsifying) and physicochemical properties of muscle proteins. Effects of heating, freezing, pH and species, *J. Food Sci.*, 50, 1034, 1985.

50. **Li-Chan, E., Kwan, L., Nakai, S., and Wood, D. F.**, Physicochemical and functional properties of salt-extractable proteins from chicken breast muscle deboned after different postmortem holding times, *Can. Inst. Food Sci. Technol. J.*, 79, 247, 1986.

51. **Kato, A. and Nakai, S.**, Hydrophobicity determined by a fluorescence probe method and its correlation with surface properties of proteins, *Biochim. Biophys. Acta*, 624, 13, 1980.

52. **Li-Chan, E.**, unpublished data, 1986.

53. **Li-Chan, E., Nakai, S., and Wood, D. F.**, Muscle protein structure-function relationships and discrimination of functionality by multivariate analysis, *J. Food Sci.*, 52, 37, 1987.

54. **Gibrat, R. and Grignon, C.**, Measurement of the quantum yield of 8-anilino-1-naphthalene-sulfonate bound on plant microsomes. Critical application of the method of Weber and Young, *Biochim. Biophys. Acta*, 691, 233, 1982.

55. **Mozhaev, V. V. and Martinek, K.**, Structure-stability relationships in proteins: new approaches to stabilizing enzymes, *Enzyme Microb. Technol.*, 6, 50, 1984.

56. **Wood, D. F., Campbell, C. A., Li-Chan, E., and Nakai, S.**, Hydrophobicity and water-holding parameters of pre- and post-rigor beef neck muscle, Abstr. 38, Can. Inst. Food Sci. Technol. 27th Ann. Conf., Vancouver, Canada, 1984.

57. **Oreshkin, E. F., Borisova, M. A., Tchubarova, G. S., Gothatov, V. M., Permyakov, E. A., Shnyrov, V. L., and Burskin, E. A.**, Conformational changes in the muscle proteins of cured beef during heating, *Meat Sci.*, 16, 297, 1986.

58. **Schmidt, G. R., Mawson, R. F., and Siegel, D. G.**, Functionality of a protein matrix in comminuted meat products, *Food Technol.*, 35(5), 235, 1981.

59. **Acton, J. C., Ziegler, G. R., and Burge, D. L., Jr.**, Functionality of muscle constituents in the processing of comminuted meat products, *CRC Crit. Rev. Food Sci. Nutr.*, 18, 99, 1983.

60. **Powrie, W. D. and Tung, M. A.**, Food dispersions, in *Principles of Food Science. Part I. Food Chemistry*, Fennema, O. R., Ed., Marcel Dekker, New York, 1976, 539.

61. **Jones, K. W.,** Protein-lipid interactions in processed meats, in Proc. 37th Ann. Reciprocal Meat Conf. American Meat Science Association, National Livestock and Meat Board, 1984, Chicago, Ill., 1985, 52.

62. **Hansen, L. J.,** Emulsion formation in finely comminuted sausage, *Food Technol.,* 14, 565, 1960.

63. **Swift, C. E., Lockett, C., and Fryar, A. J.,** Comminuted meat emulsions — the capacity of meats for emulsifying fat, *Food Technol.,* 15, 468, 1961.

64. **Yasumatsu, K., Sawada, K., Moritaka, S., Misaki, M., Toda, J., Wada, T., and Ishii, K.,** Whipping and emulsifying properties of soybean products, *Agric. Biol. Chem.,* 36, 719, 1972.

65. **Pearce, K. N. and Kinsella, J. E.,** Emulsifying properties of proteins: evaluation of a turbidimetric technique, *J. Agric. Food Chem.,* 26, 716, 1978.

66. **Ivey, F. J., Webb, N. B., and Jones, V. A.,** The effect of disperse phase droplet size and interfacial film thickness on the emulsifying capacity and stability of meat emulsions, *Food Technol.,* 24, 91, 1970.

67. **Johnson, H. C., Aberle, E. D., Forrest, J. C., Haugh, C. G., and Judge, M. D.,** Physical and chemical influences on meat emulsion stability in a model emulsitator, *J. Food Sci.,* 42, 522, 1977.

68. **Baliga, B. R. and Madaiah, N.,** Quality of sausage emulsion prepared from mutton, *J. Food Sci.,* 35, 83, 1970.

69. **Hegarty, G. R., Bratzler, L. J., and Pearson, A. M.,** Studies on the emulsifying properties of some intracellular beef muscle proteins, *J. Food Sci.,* 28, 663, 1963.

70. **Tsai, R., Cassens, R. G., and Briskey, E. J.,** The emulsifying properties of purified muscle proteins, *J. Food Sci.,* 37, 286, 1972.

71. **Kinsella, J. E.,** Functional properties of proteins in foods: a survey, *Crit. Rev. Food Sci. Nutr.,* 7, 219, 1976.

72. **Borderias, A. J., Jimenez-Colmenero, F., and Tejada, M.,** Viscosity and emulsifying ability of fish and chicken muscle protein, *J. Food Technol.,* 20, 31, 1985.

73. **Carpenter, J. A. and Saffle, R. L.,** A simple method for estimating the emulsifying capacity of various sausage meats, *J. Food Sci.,* 29, 774, 1964.

74. **Inklaar, P. A. and Fortuin, J.,** Determining the emulsifying and emulsion stabilizing capacity of protein meat additives, *Food Technol.,* 23, 103, 1969.

75. **Gabrowska, J. and Sikorski, Z.,** The emulsifying capacity of fish proteins, *Proc. 4th Int. Congr. Food Sci. and Technol.,* 11, 13, 1974.

76. **Gillett, T. A., Meiburg, D. E., Brown, C. L., and Simon, S.,** Parameters affecting meat protein extraction and interpretation of model system data for meat emulsion formation, *J. Food Sci.,* 42, 1606, 1977.

77. **Helmer, R. L. and Saffle, R. L.,** Effect of chopping temperature on the stability of sausage emulsions, *Food Technol.,* 17, 1195, 1963.

78. **Perchonok, M. H. and Regenstein, J. M.,** Stability at comminution chopping temperatures of model chicken breast muscle emulsions, *Meat Sci.,* 16, 17, 1986.

79. **Acton, J. C. and Saffle, R. L.,** Stability of oil-in-water emulsions. 1. Effects of surface tension, level of oil, viscosity and type of meat protein, *J. Food Sci.,* 35, 852, 1970.

80. **Becher, P.,** *Emulsions: Theory & Practice,* 2nd ed., Becher, P., Ed., Krieger Publishing, Melbourne, Fla., 1977.

81. **van Eerd, J.-P.,** Meat emulsion stability. Influence of hydrophilic lipophilic balance, salt concentration and blending with surfactants, *J. Food Sci.,* 36, 1121, 1971.

82. **Griffin, W. C.,** Classification of surface active agents by "HLB", *J. Soc. Cosmet. Chem.,* 1, 311, 1949.

83. **Gaska, M. T. and Regenstein, J. M.,** Timed emulsification studies with chicken breast muscle: soluble and insoluble myofibrillar proteins, *J. Food Sci.,* 47, 1438, 1982.

84. **Gaska, M. T. and Regenstein, J. M.,** Timed emulsification studies with chicken breast muscle: whole muscle, low-salt-washed muscle and low-salt soluble proteins, *J. Food Sci.,* 47, 1460, 1982.

85. **Perchonok, M. H. and Regenstein, J. M.,** Stability at cooking temperatures of model chicken breast muscle emulsions, *Meat Sci.,* 16, 31, 1986.

86. **Ziegler, G. R. and Acton, J. C.,** Heat-induced transitions in the protein-protein interaction of bovine natural actomyosin, *J. Food Biochem.,* 8, 25, 1984.

87. **Ziegler, G. R. and Acton, J. C.,** Mechanisms of gel formation by proteins of muscle tissue, *Food Technol.,* 38(5), 77, 1984.

88. **Samejima, K., Ishioroshi, M., and Yasui, T.,** Relative roles of the head and tail portions of the molecule in the heat-induced gelation of myosin, *J. Food Sci.,* 46, 1412, 1981.

89. **Hamm, R.,** Biochemistry of meat hydration, *Adv. Food Res.,* 10, 355, 1960.

90. **Wierbicki, E., Kunkle, L. E., and Deatherage, F. E.,** Changes in the water-holding capacity and cationic shifts during the heating and freezing and thawing of meat as revealed by a simple centrifugation method for measuring shrinkage, *Food Technol.,* 11(2), 69, 1957.

91. **Miller, W. O., Saffle, R. L., and Zirkle, S. B.,** Factors which influence the water-holding capacity of various types of meat, *Food Technol.,* 22, 1139, 1968.

92. **Quinn, J. R. and Paton, D.,** A practical measurement of water hydration capacity of protein materials, *Cereal Chem.,* 56, 38, 1979.

93. **Porteous, J. D. and Wood, D. F.,** Water binding of red meats in sausage formulation, *Can. Inst. Food Sci. Technol. J.,* 16, 212, 1983.
94. **Hermansson, A.-M.,** Functional properties of proteins for food: Swelling, *Lebensmitt. -Wiss. Technol.,* 5, 24, 1972.
95. **Lin, M. J. Y., Humbert, E. S., and Sosulski, F. W.,** Certain functional properties of sunflower meal products, *J. Food Sci.,* 39, 368, 1974.
96. **Wang, J. C. and Kinsella, J. E.,** Functional properties of novel proteins: alfalfa leaf protein, *J. Food Sci.,* 47, 286, 1976.
97. **Hutton, C. W. and Campbell, A. M.,** Water and fat absorption, in *Protein Functionality in Foods,* Cherry, J. P., Ed., ACS Symp. Ser. 147, American Chemical Society, Washington, D.C., 1981, 177.
98. **Voutsinas, L. P. and Nakai, S.,** A simple turbidimetric method for determining the fat binding capacity of proteins, *J. Agric. Food Chem.,* 31, 58, 1983.
99. **Meyer, J. A., Brown, W. L., Giltner, M. E., and Guinn, J. R.,** Effect of emulsifiers on the stability of sausage emulsions, *Food Technol.,* 18, 1976, 1964.
100. **Brown, D. D. and Toledo, R. T.,** Relationship between chopping temperature and fat and water binding in comminuted meat batters, *J. Food Sci.,* 40, 1061, 1975.
101. **Hamm, R.,** Post-mortem changes in muscle affecting the quality of comminuted meat products, in *Developments in Meat Science. 2.,* Lawrie, R., Ed., Elsevier, New York, 1981, 93.
102. **Swift, C. E. and Berman, M. D.,** Factors affecting the water retention of beef. 1. Variations in composition and properties among eight muscles, *Food Technol.,* 13, 365, 1959.
103. **Mittal, G. S. and Blaisdell, J. L.,** Weight loss in frankfurters during thermal processing, *Meat Sci.,* 9, 79, 1983.
104. **Torgersen, H. and Toledo, R. T.,** Physical properties of protein preparations related to their functional characteristics in comminuted meat systems, *J. Food Sci.,* 42, 1615, 1977.
105. **Offer, G. and Trinick, J.,** On the mechanism of water holding in meat: the swelling and shrinking of myofibrils, *Meat Sci.,* 8, 245, 1983.
106. **Johnson, P. G. and Bowers, J. A.,** Influence of aging on the electrophoretic and structural characteristics of turkey breast muscle, *J. Food Sci.,* 41, 255, 1976.
107. **Matsumoto, J. J.,** Chemical deterioration of muscle proteins during frozen storage, in *Chemical Deterioration of Proteins,* Whitaker, J. R. and Fujimaki, M., Eds., ACS Symp. Ser. 123, American Chemical Society, Washington, D.C., 1980, 95.
108. **Shenouda, S. Y. K.,** Theories of protein denaturation during frozen storage of fish flesh, *Adv. Food Res.,* 26, 275, 1980.
109. **Dyer, W. J. and Dingle, J. R.,** Fish proteins with special reference to freezing, in *Fish as Food,* Borgstrom, G., Ed., Academic Press, New York, 1961, 275.
110. **Sikorski, Z., Olley, J., and Kostuch, S.,** Protein changes in frozen fish, *CRC Crit. Rev. Food Sci. Nutr.,* 8, 97, 1976.
111. **Shenouda, S. Y. K. and Pigott, G. M.,** Lipid protein interaction during aqueous extraction of fish protein: actin-lipid interaction, *J. Food Sci.,* 40, 523, 1975.
112. **Shenouda, S. Y. K. and Pigott, G. M.,** Fish myofibrillar protein and lipid interaction in aqueous media as detected by isotope labeling, sucrose gradient centrifugation, polyacrylamide electrophoresis and electron paramagnetic resonance, in *Protein Crosslinking: Biochemical and Molecular Aspects,* Friedman, M., Ed., *Adv. Exper. Med. Biol.* 86-A, 1977, 657.
113. **Shenouda, S. Y. K. and Pigott, G. M.,** Electron paramagnetic resonance studies of actin-lipid interaction in aqueous media, *J. Agric. Food Chem.,* 24, 11, 1976.
114. **Shenouda, S. Y. K. and Pigott, G. M.,** Lipid protein interaction during aqueous extraction of fish protein. Myosin-lipid interaction, *J. Food Sci.,* 39, 726, 1974.
115. **Wu, M. C., Akahane, T., Lanier, T. C., and Hamann, D. D.,** Thermal transitions of actomyosin and surimi prepared from Atlantic croaker as studied by differential scanning calorimetry, *J. Food Sci.,* 50, 10, 1985.
116. **Wu, M. C., Lanier, T. C., and Hamann, D. D.,** Rigidity and viscosity changes of croaker actomyosin during thermal gelation, *J. Food Sci.,* 50, 14, 1985.
117. **Hamm, R.,** Changes of muscle proteins during the heating of meat, in *Physical, Chemical and Biological Changes in Food Caused by Thermal Processing,* Hoyem, T. and Kvale, O., Eds., Applied Science Publishers, London, 1977, 101.
118. **Acton, J. C. and Dick, R. L.,** Protein-protein interaction in processed meats, Proc. 37th Ann. Reciprocal Meat Conf. American Meat Science Association, National Livestock and Meat Board, 1984, Chicago, Ill., 1985, 36.
119. **Acton, J. C., Hanna, M. A., and Satterlee, L. D.,** Heat-induced gelation and protein-protein interaction of actomyosin, *J. Food Biochem.,* 5, 101, 1981.
120. **Wright, D. J., Leach, I. B., and Wilding, P.,** Differential scanning calorimetric studies of muscle and its constituent proteins, *J. Sci. Food Agric.,* 28, 557, 1977.

121. **Wright, D. J. and Wilding, P.**, Differential scanning calorimetric study of muscle and its proteins: Myosin and its subfragments, *J. Sci. Food Agric.*, 3, 457, 1984.

122. **Stabursvik, E. and Martens, H.**, Thermal denaturation of proteins in post-rigor muscle tissue as studied by differential scanning calorimetry, *J. Sci. Food Agric.*, 31, 1034, 1980.

123. **Samejima, K., Takahashi, K., and Yasui, T.**, Heat-induced denaturation of myosin total rod, *Agric. Biol. Chem.*, 40, 2455, 1976.

124. **Hamm, R.**, Heating of muscle systems, in *The Physiology and Biochemistry of Muscle as a Food*, Briskey, E. J., Cassens, R. G., and Trautman, J. C., Eds., University of Wisconsin Press, Madison, 1966, 363.

125. **Ishioroshi, M., Samejima, K., and Yasui, T.**, Further studies on the roles of the head and tail regions of the myosin molecule in heat-induced gelation, *J. Food Sci.*, 47, 114, 1982.

126. **Liu, Y. M., Lin, T. S., and Lanier, T. C.**, Thermal denaturation and aggregation of actomyosin from Atlantic croaker, *J. Food Sci.*, 47, 1916, 1982.

127. **Wicker, L., Lanier, T.C., Hamann, D.D., and Akahane, T.**, Thermal transitions in myosin-ANS fluorescence and gel rigidity, *J. Food Sci.*, 51, 1540, 1986.

128. **Fukazawa, T., Hashimoto, Y., and Yasui, T.**, Effects of some proteins on the binding quality of an experimental sausage, *J. Food Sci.*, 26, 541, 1961.

129. **Goodno, C. C. and Swenson, C. A.**, Thermal transitions of myosin and its helical fragments. I. Shifts in proton equilibria accompanying unfolding, *Biochemistry*, 14, 867, 1975.

130. **Goodno, C. C. and Swenson, C. A.**, Thermal transitions of myosin and its helical fragments. II. Shifts in proton equilibria accompanying unfolding, *Biochemistry*, 14, 873, 1975.

131. **Deng, J., Toledo, R. T., and Lilliard, D. A.**, Effect of temperature and pH on protein-protein interaction in actomyosin solutions, *J. Food Sci.*, 41, 273, 1976.

132. **Siegel, D. G. and Schmidt, G. R.**, Crude myosin fractions as binders, *J. Food Sci.*, 44, 1129, 1979.

133. **Siegel, D. G. and Schmidt, G. R.**, Ionic, pH and temperature effects on the binding ability of myosin, *J. Food Sci.*, 44, 1686, 1979.

134. **Yasui, T., Ishioroshi, M., Nakano, H., and Samejima, K.**, Changes in shear modulus, ultrastructure and spin-spin relaxation times of water associated with heat-induced gelation of myosin, *J. Food Sci.*, 44, 1201, 1979.

135. **Yasui, T., Ishioroshi, M., and Samejima, K.**, Effect of actomyosin on heat-induced gelation of myosin, *Agric. Biol. Chem.*, 46, 1049, 1982.

136. **Ishioroshi, M., Samejima, K., and Yasui, T.**, Heat-induced gelation of myosin: factors of pH and salt concentrations, *J. Food Sci.*, 44, 1280, 1979.

137. **Samejima, K., Hashimoto, Y., Yasui, T., and Fukazawa, Y.**, Heat gelling properties of myosin, actomyosin and myosin subunits in a saline model system, *J. Food Sci.*, 34, 242, 1969.

138. **Samejima, K., Yamauchi, H., Ashgar, A., and Yasui, T.**, Role of myosin heavy chains from rabbit skeletal muscle in the heat-induced gelation mechanism, *Agric. Biol. Chem.*, 48, 2225, 1984.

139. **Samejima, K., Egelandsal, B., and Fretheim, K.**, Heat gelation properties and protein extractability of beef myofibrils, *J. Food Sci.*, 50, 1540, 1985.

140. **Montejano, J. G., Hamann, D. D. and Lanier, T. C.**, Thermally induced gelation of selected comminuted muscle systems — rheological changes during processing, final strengths and microstructure, *J. Food Sci.*, 49, 1496, 1984.

141. **Montejano, J. G., Hamann, D. D., and Lanier, T. C.**, Final strengths and rheological changes during processing of thermally induced fish muscle gels, *J. Rheol.*, 27, 557, 1983.

142. **Lanier, T. C.**, Fabricated seafood products. Functional properties of surimi, *Food Technol.*, 40(3), 107, 1986.

143. **Lanier, T. C., Lin, T. S., Liu, Y. M., and Hamann, D. D.**, Heat gelation properties of actomyosin and surimi prepared from Atlantic croaker, *J. Food Sci.*, 47, 1921, 1982.

144. **Niwa, E., Nakayama, T., and Hamada, I.**, Attempt to determine the state of water in fish gel by IR technique, *Bull. Jpn. Soc. Sci. Fish.*, 46, 863, 1980.

145. **Niwa, E., Nakayama, T., and Hamada, I.**, Preparation of myosin heavy chain from fish muscle, *Bull. Jpn. Soc. Sci. Fish.*, 46, 867, 1980.

146. **Niwa, E. and Nakajima, G.**, Differences in protein structure between elastic kamaboko and brittle one, *Bull. Jpn. Soc. Sci. Fish.*, 41, 579, 1975.

147. **Ueda, T., Simidu, W., and Shimizu, Y.**, Studies on muscles of aquatic animal. XXXVIII. Change in viscosity of heat denatured fish actomyosin, *Bull. Jpn. Soc. Sci. Fish.*, 29, 537, 1963.

148. **Ueda, T., Shimizu, Y., and Simidu, W.**, Studies on muscle of aquatic animals. XXXX. Species difference in fish actomyosin, Part I, Relationship between the viscosity and ionic strength, *Bull. Jpn. Soc. Sci. Fish.*, 29, 794, 1963.

149. **Ueda, T., Shimizu, Y., and Simidu, W.**, Studies on muscle of aquatic animals. XXXXII. Species differences in fish actomyosin. 2. Relationship between heat denaturing point and species, *Bull. Jpn. Soc. Sci. Fish.*, 31, 352, 1964.

150. **Ueda, T., Shimizu, Y., and Simidu, W.,** Species differences — fish muscle. 1. The gel-forming ability of heated ground muscles, *Bull. Jpn. Soc. Sci. Fish.*, 34, 357, 1968.

151. **Shimizu, Y. Nishioka, F., Machida, R., and Shiue, C. M.,** Gelation characteristics of myosin sol with added salt, *Bull. Jpn. Soc. Sci. Fish.*, 49, 1239, 1984.

152. **Niwa, E., Nakayama, T., and Hamada, I.,** The third evidence for the participation of hydrophobic interactions in fish flesh gel formation, *Bull. Jpn. Soc. Sci. Fish.*, 49, 1763, 1983.

153. **Egelansdal, B., Freithem, K., and Harbitz, O.,** Fatty acid salts and analogs reduce thermal stability and improve gel formability of myosin, *J. Food Sci.*, 50, 1399, 1985.

154. **Taguchi, T., Tanaka, M., and Suzuki, K.,** "Himidori" (thermally induced disintegration) of oval filefish myosin gel, *Bull. Jpn. Soc. Sci. Fish.*, 49, 1281, 1983.

155. **Swift, C. E. and Sulzbacher, W. L.,** Factors affecting meat proteins as emulsion stabilizers, *Food Technol.*, 17, 106, 1963.

156. Computer Concepts Corporation, Computer least cost formulation, *Food Processing*, 36, 74, 1975.

157. **Kramlich, W. E.,** Sausage products, in *The Science of Meat and Meat Products*, Price, J. F. and Schweigert, B. S., Eds., W. H. Freeman, San Francisco, Calif., 1971, 484.

158. **Anderson, H. V. and Clifton, E. S.,** How the small plant can profitably use least cost sausage formulation, *Meat Processing*, 2, 7, 1967.

159. **Brown, D. D.,** A study of factors affecting stability and quality, Ph.D. thesis, University of Georgia, Athens, 1972, cited by Gillett, T. A., Meiburg, D. E., Brown, C. L., and Simon, S., *J. Food Sci.*, 42, 1606, 1977.

160. **Regenstein, J. M.,** Protein-water interactions in muscle foods, in Proc. 37th Ann. Reciprocal Meat Conf. American Meat Science Association, National Live Stock and Meat Board, 1984, Chicago, Ill., 1985, 44.

Chapter 5

IMPORTANCE OF HYDROPHOBIC INTERACTIONS IN MODIFICATION OF STRUCTURE AND FUNCTION OF FOOD PROTEINS

TABLE OF CONTENTS

I. INTRODUCTION

Both intrinsic and extrinsic factors affect the functional properties of protein. In the former category are the basic physicochemical properties of the proteins, including their steric, electric, and hydrophobic/hydrophilic characteristics. In the latter category are the environment of the protein product (including pH, ionic strength, and other food components such as carbohydrates, lipids, minerals, salts, etc.), and processing and modification steps which alter the intrinsic protein properties as well as their environment.

According to Pour-El[1], there are seven major process steps which commonly influence functionality: protein source and variety, extraction, temperature, drying, ionic history, impurities, and storage. Three main avenues are pursued for altering the functional properties of a protein product, namely physical, chemical, and biological methods. Thus, modification of food proteins can involve alterations in the protein structure or conformation at all levels of organization, i.e., primary, secondary, tertiary, and quaternary structures. Physical methods include processes such as heating, fiber spinning, extrusion, and sonication, while chemical and biological methods may include enzymatic hydrolysis, synthesis, or resynthesis techniques, chemical derivatization, and acid or alkali hydrolysis. In addition, modification may include complex formation with other food components or additives such as flavor compounds, tannins, lipids, surfactants, carbohydrates, and other proteins.

Chemical, enzymatic, and physical modifications thus offer great potential in improving and extending the use of food proteins from both conventional and novel sources. Feeney[2,3] has listed some of the possible applications of chemical modification. However, the intentional chemical modification of food proteins is currently very limited, partly due to cultural, socioeconomic, and legal considerations, but also most likely due to inability to predict properties of the modified proteins. Even in the case of biological and physical processes for modification, a general reluctance to change current technology is prevalent.

Two basic issues on enhancement of protein utilization in foods through functionality improvement were addressed by Ryan[4]. Firstly, what are the limitations in our current knowledge which are holding back the development of methodologies for alteration of specific functional properties, and secondly, what kind of general conceptual framework would allow clear understanding of protein functionality and rapid application for intentional control and improvement of protein functionality? In response to the first issue, Ryan[4] suggested that the largest obstacle in development of methodologies for functionality alteration is our present inability to predict how a particular chemical or structural change will affect a specific functional property. In response to the second question on the approach to understanding functionality, detailed understanding of both protein structure and intermolecular interactions are needed as the initial step to correlating structure and functionality. Unless protein structure can be meaningfully correlated with specific functional properties, the design of processes to intentionally modify functional properties will probably have to continue to rely on a trial-and-error basis, with a high failure rate.

Over the last decade, food scientists have been recognizing the importance of correlating protein structure to functionality as a systematic and more generally versatile approach for understanding and designing process and modification methodology to improve functionality. A number of reviews and symposia have been published recently in this respect.[5-11] This basic information is also crucial for food scientists to meet the challenges of protein engineering and biotechnology.

In this chapter, the role of hydrophobic interactions in protein functionality and their alteration through various modification techniques and processes are emphasized. However, the reader should keep in mind that in addition to hydrophobic interactions, a variety of other molecular interactions may also be altered simultaneously, including other noncovalent forces as well as covalent bonds, and that the realization of the interplay between these

Table 1
CHEMICAL MODIFICATION OF AMINO
ACID SIDE CHAINS

Side chain	Amino acid residue	Commonly used modifications
Amino	Lysine	Alkylation, acylation
Carboxyl	Aspartic and glutamic acid	Esterification, amide formation
Disulfide	Cystine	Reduction, oxidation
Sulfhydryl	Cysteine	Alkylation, oxidation
Thioether	Methionine	Alkylation, oxidation
Imidazole	Histidine	Oxidation, alkylation
Indole	Tryptophan	Oxidation, alkylation
Phenolic	Tyrosine	Acylation, electrophilic substitution
Guanidino	Arginine	Condensation with dicarbonyls

interactions is important in assessing the relationship between structure modification and functionality.

II. CHEMICAL MODIFICATION

Chemical modification of proteins has been widely used in basic protein chemistry to study the relationships between structure and biological function.[12-15] More recently, application of chemical modification techniques to probe the structure of food proteins as well as to improve their nutritional and functional properties has been an area of expanding interest to food protein chemists, as illustrated by several excellent reviews.[2-4,9,10,16-22] Some of the more commonly used methods of chemical modification of amino acid residues are shown in Table 1.

The majority of studies on food protein structure-function have involved derivatization of the epsilon amino group of lysine residues, although modification of ω-carboxyl groups of aspartyl and glutamyl residues has also been investigated. These types of derivatization can directly affect the net charge and charge-density of the protein molecules, but in addition they often result in conformational changes and alterations in intra- and intermolecular interactions which also modify effective hydrophobicity of the proteins.

A. Acylation

Acylation of amino acid residues, particularly lysine residues, has probably been the most common chemical derivatization used in food protein applications. Acetylation (typically using acetic anhydride) replaces the positively charged epsilon amino groups of lysine residues by neutral acetyl groups, while succinylation (typically using succinyl anhydride) changes the positively charged groups to negatively charged succinyl anionic groups. Generally, although acetylation increases solubility of food proteins, enhancement of functional properties has not always been observed.[18] In addition to the slight increase in aqueous solubility, acetylation often decreases the isoelectric point due to reduced positive charges, and some decrease in heat-induced gelation may be observed. On the other hand, succinylation brings about larger changes in electrostatic nature of the proteins due to the elimination of some positively charged groups in conjunction with introduction of the negatively charged succinyl groups and thus, generally large changes in physicochemical as well as functional properties of succinylated proteins have been observed. Alterations in functional properties of succinylated proteins have commonly included increased aqueous solubility, altered viscosity, enhanced hydration or wettability, and modified surfactant properties such as emulsifying and foaming abilities.[17,18] Increased resistance to aggregation or thermal instability has also been noted.[18,19]

Proteins that have been acylated to alter functionality include yeast protein,[18] casein,[23,24] whey proteins,[25] fish proteins,[26-28] single cell protein,[29] soy proteins,[30] sunflower protein,[31] leaf protein concentrate,[32] pea protein isolate,[33] cottonseed protein,[34] wheat flour protein,[35] egg white proteins,[36-41] peanut protein,[42] oat protein,[43] and muscle proteins.[44,45] Changes in the acylated protein functionality are hypothesized to be due not only to improved solubility properties caused by altered charge properties, but also due to loosening and unfolding of the acylated protein structure; this unfolding simultaneously exposes polar and nonpolar residues which can contribute to the improvement in various functional properties.

For example, Habeeb et al.[46] reported that succinylated proteins such as β-lactoglobulin and bovine serum albumin exhibited increased intrinsic viscosities with concomitant decrease in sedimentation coefficients at neutral and alkaline pH, arising from molecular unfolding and expansion. Hollecker and Creighton[47] studied the effect of varying extents of succinylation by urea-gradient electrophoresis. For cytochrome c and ribonuclease, net stabilities of their folded states were varied only slightly by succinylation. For β-lactoglobulin, reaction of the initial amino groups produced a small increase in stability in a few instances and a decrease in others; succinylation of more than ten lysyl residues abruptly caused unfolding in the absence of urea. Shetty and Rao[42] reported dissociation of succinylated arachin (peanut) protein into low molecular-weight components was suggested from sedimentation patterns; yet an increase in viscosity was observed. This suggested that unfolding of the dissociated components and the resulting increase in hydrodynamic volume more than compensated for the decrease in viscosity which would be expected to accompany dissociation. Progressive succinylation of the arachin protein caused conformational changes and polypeptide unfolding, as indicated by a blue shift of the ultraviolet absorption spectrum at 287 nm corresponding to the tyrosine peak, and an increased specific ellipticity of the dichroic spectra at 212 nm. Gray and Lomath[48] reported that, like succinylation, citraconylation of human serum albumin caused a significant increase in apparent Stokes radius and viscosity; however, no significant change in % helix content was measured by optical rotatory dispersion and it was postulated that the apparent volume increase may have been due to an enlarged solvation shell rather than to major unfolding of the molecule. Nevertheless, enhancement in thermal stability of citraconylated albumin was observed, even after prolonged exposure to high temperature which resulted in removal of most citraconylated groups, suggesting that some structural changes had been induced by citraconylation.

Alterations in intermolecular interactions and aggregation may arise from acylation. Hoagland and co-workers[49,50] and Evans et al.[51,52] studied the effects of alkyl group size on the calcium ion sensitivity and aggregation properties of acylated β-casein. Although succinylation hindered association of β-casein molecules, acylation with a series of anhydrides produced derivatives which showed enhanced aggregation with increasing length of the substituent n-alkyl chain. Thus, acetyl and propionyl β-caseins formed monomer-polymer systems similar to native β-casein, but associated to a lesser extent probably due to an increase negative repulsive forces. On the other hand, n-hexanoyl, n-octanoyl, and n-decanoyl derivatives associated strongly, and the proportion of polymers increased with increasing length of n-alkyl chain substituent, suggesting the presence of hydrophobic bonding. The higher acyl β-casein derivatives contained mixtures of α-helix and random coil structures as measured by optical rotatory dispersion, while nuclear magnetic resonance spectroscopy revealed side-chain interactions which were not present in the native protein. It was suggested that structural organization of aggregates of the derivatives included location of protein peptide bonds in strongly hydrophobic environments.

A unique type of acylation is that in which amino groups are thiolated using S-acetylmercaptosuccinic anhydride (S-ASMA) or *N*-acetylhomocysteine thiolactone (N-AHTL), which effectively introduces new sulfhydryl groups onto the modified protein.[17,53] Oxidation of thiolated β-lactoglobulin produced novel polymers exhibiting unique functional characteristics and enhanced thermal stability. Similarly, introduction of new thiol groups on to soy protein for functionality improvement was accomplished enzymatically by papain-catalyzed acylation between *N*-AHTL and amino groups of soy protein.[54] Unfortunately, the role of hydrophobic interactions and other noncovalent interactions in the structure and function of these thiolated proteins has not been investigated.

B. Esterification/Amidation

Carboxyl groups of aspartate and glutamate residues of proteins can be blocked by amidation or esterification reactions, which effectively reduces the net negative charge or conversely increases the net positive charge of the protein, with accompanying changes in physicochemical and functional properties.

Esterification of carboxyl groups may be accomplished by suspending the protein in the appropriate alcohol with an acid catalyst.[55-58] In practice, methyl and ethyl esters of protein carboxyl groups are formed readily. In the case of β-lactoglobulin, for example, esterification in methanol or ethanol occurred over a period of several days at 4°C, with increasing solubilization of the protein in the alcohol as esterification progressed.[21] Following completion of the esterification process, excess alcohol could be easily removed by dialysis. Although longer chain esters are difficult to prepare directly, it is likely that acid-catalyzed transesterification of protein methyl esters in higher alcohols could be used for indirect preparation of the longer chain esters to yield more hydrophobic proteins.[21] Amidation of carboxyl groups may be accomplished via a carbodiimide-mediated condensation of carboxyl groups with ammonium ion, yielding asparaginyl and glutaminyl residues from aspartate and glutamate residues, respectively,[57-59] or with esters of amino acids.[39]

Effects of amidation or esterification of bovine β-lactoglobulin were studied by Mattarella and co-workers.[56-58] A more random structure indicative of partial denaturation as a result of amidation or esterification was evident by circular dichroism measurements. Thus, the structure of native β-lactoglobulin was calculated from circular dichroism data as being 16% α-helical, 54% β-sheet, and 30% aperiodic or random coil structure, whereas the structure of amidated β-lactoglobulin was about 20% α-helical, 29% β-sheet and 51% aperiodic.[57] Changes in conformation of amidated or esterified derivatives were also suggested by disappearance of troughs in the near-ultraviolet circular dichroism spectra, which are usually indicative of asymmetric energy transfer between aromatic chromophores. Ultraviolet difference absorbance spectra also indicated that aromatic residues of the modified proteins were in a different environment compared to the native protein.[58]

Surface hydrophobicities of amidated and esterified β-lactoglobulin derivatives were compared to the native protein using fluorescence probes (*cis*-parinarate or CPA and anilinonaphthalene sulfonate or ANS) as well as heptane-binding measurements. Some differences were noted between the three hydrophobicity parameters. The following orders of decreasing hydrophobicity were observed: (1) by heptane binding: amidated, ethyl-esterified, methylesterified, and native; (2) by ANS fluorescence probe: methyl-esterified, ethyl-esterified, amidated, and native; and (3) by CPA fluorescence probe: methyl-esterified, amidated, ethylesterified and native. Mattarella et al.[58] suggested that the differences may arise from the charged nature of the fluorescence probes which may not bind merely through hydrophobic interactions. However, it is also possible that the differences arose from differences in exposure of the two types of hydrophobicity, aromatic and aliphatic, as suggested by Hayakawa and Nakai,[60] and from the effects of varying ionic strength which may indeed reflect changes in protein conformation. Nevertheless, despite these differences (Table 2), in general it could be concluded that the esterified as well as amidated proteins had greater hydropho-

Table 2
**HYDROPHOBIC PARAMETERS OF NATIVE,
AMIDATED AND ESTERIFIED β-
LACTOGLOBULIN MEASURED BY *CIS*-
PARINARIC ACID (CPA) FLUORESCENCE PROBE
AND HEPTANE BINDING MEASUREMENTS**

β-Lactoglobulin sample	S_o CPA			Heptane binding (relative scale)
	No. NaCl	0.55M NaCl	2.0M NaCl	
Native	57	144	584	0.35
Ethyl-esterified	132	168	260	0.77
Methyl-esterified	420	407	225	0.46
Amidated	384	360	158	1.00

Adapted from Mattarella and Richardson.[58]

Table 3
**SURFACE AND INTERFACIAL TENSION OF
NATIVE AND ESTERIFIED β-
LACTOGLOBULIN (0.2% W/V PROTEIN
SOLUTIONS)**

β-Lactoglobulin sample	Surface tension[a] (dyne cm^{-1})	Interfacial tension[b] (dyne cm^{-1})
Native	64.4 ± 0.4	20.0 ± 0.8
Methyl-esterified	50.3 ± 0.4	24.5 ± 0.5
Ethyl-esterified	54.3 ± 0.4	16.7 ± 0.2
Butyl-esterified	56.4 ± 0.3	17.8 ± 0.3

[a] Surface tension of water was 73.6 ± 0.7 dyne cm^{-1}.
[b] Interfacial tension of corn oil-water was 33.2 ± 0.9 dyne cm^{-1}.

Adapted from Richardson.[21]

bicity properties than the native β-lactoglobulin. These increases in hydrophobicity were postulated to be related to randomization or unfolding of the protein structure as well as introduction of hydrophobic aliphatic groups by amidation and esterification.

The physicochemical changes in amidated or esterified derivatives of β-lactoglobulin were also accompanied by modified functionality. Amidated and esterified proteins had decreased solubility at low ionic strength at pH 8 to 10, close to their isoionic point; at higher ionic strength ($\mu = 0.55$) methyl-esterified derivatives were <5% and ethyl-esterified derivatives were only about 50% as soluble as native β-lactoglobulin, over the pH range from 3 to 10. Tryptic hydrolysis rate was markedly increased for the modified derivatives, partly due to their denaturation and increased randomization. Emulsifying activity of modified derivatives was slightly lower than that of the native protein, but emulsifying stability of ethyl-esterified protein-prepared emulsions was much higher than for emulsions prepared with native protein. Over 40% of the ethyl-esterified protein was adsorbed to the oil-water interface, and this unique interfacial behavior of the ethyl-esterified protein was postulated to be due to its lower isoelectric point and the increased hydrophobicity resulting from incorporation of ethyl groups. Enhanced surface activities were demonstrated from surface and interfacial tension measurements (Table 3). Although part of the enhanced surface properties of the esterified proteins may be attributed to introduction of alcohol residues on to the protein, it is likely that the major contribution to the surface activities arose from conformational changes and net surface charge alterations.[21]

Strong interaction was demonstrated between casein micelles of bovine milk and the positively charged amidated or esterified β-lactoglobulin derivatives.[56] It was estimated that addition of 1 to 2g of the derivatives (1, 2, and 1g of amidated, ethyl-esterified, and methyl-esterified derivatives, respectively) would coagulate the casein micelles in 100 mℓ of milk, without any need for addition of rennet extract.

Ma and Nakai[61,62] studied the effects of carboxyl modification of porcine pepsin by carbodiimide-mediated amide formation on its properties as well as interactions with milk and caseins. Modification of up to 11 carboxyl groups with glycine methyl ester caused changes in activities, specificity and physicochemical properties. Although caseinolytic properties of the carboxyl-modified pepsin were not affected, clotting activity against κ-casein was increased while clotting activity against κ-α_{s1}-casein mixture was decreased. The modified pepsin also possessed increased stability near neutral pH, as well as increased thermal stability with respect to milk-clotting activity, suggesting it may be a more suitable rennet substitute than native pepsin for cheese manufacturing. Ma and Nakai[62] concluded that these results showed the possibility of changing the activity, specificity, and physicochemical properties of an enzyme by chemical modification.

When methyl esters of amino acids other than glycine were used as nucleophiles for pepsin modification, milk-clotting activities of all modified enzymes were markedly reduced, particularly those modified with tyrosine and tryptophan methyl esters. At pH 2.0, proteolytic activity of pepsin modified with arginine and lysine methyl esters was unchanged, whereas pepsin modified with leucine, tyrosine, and tryptophan methyl esters showed a decrease in proteolytic activity. At pH 3.5, proteolytic activity was increased in most modified enzymes except for those modified with tyrosine and tryptophan methyl esters. It was postulated that incorporation of hydrophobic amino acids such as tyrosine and tryptophan may have decreased solubility of the pepsin, resulting in decreased activity. Alternatively, loss in both milk clotting and proteolytic activities could be due to a greater affinity of hydrophobic esters to the hydrophobic binding sites of pepsin, leading to modification of carboxyl groups near the active site of pepsin.[61]

The properties of egg albumen modified by succinylation or by carboxyl modification via carbodiimide-mediated amidation using glycine methyl ester were compared to the unmodified egg albumen, before and after heating (Table 4).[39] The unheated chemically modified proteins showed only minor conformational changes compared to the native egg albumen. There was a progressive decrease in pI with increasing extent of succinylation, and an increase in pI with carboxyl modification. However, no significant alterations in size or configuration of protein molecules were noticed, based on sedimentation velocity rates and surface hydrophobicity using a cis-parinarate probe and optical rotation measurements.

Heat treatment at 100°C for 3 min did cause significant changes in physicochemical properties of egg albumen, and these heat-induced changes were influenced by the two chemical modifications (Table 4). In particular, heat-induced coagulation or gelation was retarded by both modifications, suggesting that thermocoagulation required a balanced electrostatic attraction between protein molecules. The soluble fraction of heat-coagulated protein samples contained mainly monomers, whereas under conditions in which no gel or coagulum was formed, heat-treated albumen contained high molecular weight-soluble aggregates, as evidenced by the polymers with $S_{20,w}$ of 11.5S to 21.5S (Table 4).

A dramatic increase in surface hydrophobicity was observed in both native and modified egg albumen upon heating. Samples containing insoluble aggregates (i.e., forming solid gels) had significantly lower hydrophobicity than samples remaining as clear solutions. These results suggest that the formation of the gel or coagulum from heated albumen involved

Table 4

PROPERTIES OF NATIVE AND MODIFIED EGG ALBUMEN BEFORE AND AFTER HEAT TREATMENT (100°C, 3 MIN)

		Egg albumen			
			S-EA[a]		
	Native	**100:1**	**50:1**	**10:1**	**COOH-EA[b]**
% Modification	0	27	45	100	35
pI	4.65	4.40	4.15	3.70	4.90
SH content, μM/g					
Unheated	43.9 ± 2.1	44.4 ± 2.3	42.9 ± 1.9	40.8 ± 2.1	40.7 ± 1.7
Heated, pH 5.5	46.9 ± 2.4	39.5 ± 2.0	41.3 ± 1.8	39.9 ± 1.7	37.8 ± 1.8
Heated, pH 8.0	38.8 ± 1.2	35.2 ± 1.7	39.5 ± 2.1	40.6 ± 1.5	39.2 ± 1.6
$S_{20,w}$					
Unheated	3.52	3.54	3.54	3.56	3.52
Heated, pH 5.5	—	—	—	3.52, 21.5	3.52, 21.2
Heated, pH 8.0	3.52, 21.4	3.52, 21.5	3.52, 11.5	3.52	—
Hydrophobicity					
Unheated	10 ± 2	10 ± 2	12 ± 3	12 ± 2	10 ± 1
Heated, pH 5.5	208 ± 12	220 ± 20	203 ± 15	770 ± 40	720 ± 30
Heated, pH 8.0	495 ± 25	525 ± 25	463 ± 27	392 ± 20	470 ± 25
$[\alpha]_{589}^{25}$					
Unheated	− 30.7	− 29.8	− 30.9	− 35.5	− 30.8
Heated, pH 5.5	—	—	—	− 78.9	− 62.8
Heated, pH 8.0	− 68.7	− 74.6	− 74.3	− 77.2	—
Visual appearance					
Heated, pH 5.5	Opaque solid gel	Opaque solid gel	Turbid loose gel	Clear solution	Clear solution
Heated, pH 8.0	Turbid solution	Clear solution	Clear solution	Clear solution	Turbid loose gel

[a]   S-EA = succinylated egg albumen with the indicated protein: succinic anhydride ratio (w:w).

[b]   COOH-EA = carboxyl-modified egg albumen.

Data compiled from Ma and Holme.[39]

direct interaction of hydrophobic sites between heat-denatured protein molecules, resulting in a decrease in exposed hydrophobic sites after interaction. In contrast, the high hydrophobicity of the extensively succinylated (10:1) and the carboxyl-modified heated derivatives suggest that these modifications prevented intermolecular hydrophobic interactions, resulting in inability to form solid gels.

C. Reductive Alkylation

Reductive alkylation of the amino groups of proteins may be achieved by reacting the protein with an aldehyde or ketone in the presence of a reducing agent of the hydride donor type, such as sodium borohydride, cyanoborohydride, or amino boranes. Using formaldehyde as the carbonyl component, the di-substituted derivatives with ϵ-N,N-dimethyllysine residues are the principal products, whereas with other aldehydes and ketones, predominantly monoalkyllysine residues are formed, the reaction with a second carbonyl molecule being greatly retarded.[13]

Alkylation does not greatly alter the basicity of amino groups of proteins, the pK values of dimethylamino groups (tertiary amines) being about 0.4 to 0.6 pH units below those of primary amino groups, and the pK values of monoalkylamino groups (secondary amines) being about 0.1 to 0.7 pH units above those of primary amino groups.[13] The relatively small effect on pK of the derivatized amino group and the small size of methyl groups usually result in only minimal changes to proteins which have been modified by reductive methylation.[63] For example, reductive methylation of α_s-, β-, and κ-caseins had only minor effects on their physicochemical characteristics such as solubility, electrophoretic mobility, and susceptibility to calcium-induced precipitation.[64] Similarly, reductive methylation of up to 80% of the amino groups of β-lactoglobulin did not result in large changes in electrophoretic mobility, isoelectric point, or heat-denaturation profile.[65] However, subtle changes in conformation of methylated proteins have been reported.[17]

By using different sized substituents on the carbonyl compound, the effects of addition of groups of differing hydrophobicity on protein conformation and properties may be varied. For example, Fretheim et al.[66] alkylated ovomucoid, lysozyme, and ovotransferrin with various carbonyl reagents in the presence of sodium borohydride. The methylated and isopropylated derivatives of all three proteins were soluble and they retained almost full biochemical activities, but introduction of larger substituents caused precipitation of lysozyme and ovotransferrin.

Sen et al.[67] reported that the conformation of casein was altered by reductive alkylation to an extent which depended on the size of the hydrophobic alkyl group. Six alkylated derivatives were prepared, including methyl, isopropyl, butyl, cyclopentyl, cyclohexyl, and benzyl casein. It was suggested that highly alkylated caseins (15 or 16 alkylated residues per mol) were folded more compactly to allow for hydrophobic interactions between alkyl groups resulting in a more compact globular structure, in contrast to native casein which has very little tertiary structure. The changes in structure were manifested in differences in the ultraviolet spectra of alkylated derivatives compared to native casein; in general, the extent of deviation of the ultraviolet spectrum from that of native casein correlated with the extent of modification and/or size of the alkyl group. Significantly lower rates of α-chymotrypsin catalyzed hydrolysis of alkylated caseins as well as their lower viscosities compared to unmodified casein were attributed to their more folded, compact structure. Solubilities of methyl and isopropyl casein were slightly higher than native casein, whereas caseins alkylated with bulkier groups had lower solubility. Except for butyl and cyclopentyl caseins, all highly alkylated caseins had superior water-binding (absorbed moisture) properties than native casein, and except for butyl casein, the highly alkylated caseins all had higher emulsifying activity than native casein (Table 5). These results demonstrate the potential of reductive alkylation as a means of covalent attachment of hydrophobic substituents of varying size and structure to proteins, in order to improve properties including solubility, water binding, and emulsifying ability.

Reductive alkylation may also be used to alter properties via attachment of sugar residues. Lee et al.[68] attached glucose, fructose, or lactose to lysyl residues of casein by reductive alkylation in the presence of sodium cyanoborohydride. Although it was speculated that attachment of hydrophilic groups, such as carbohydrates to proteins, would change protein solubility, viscosity, hydration, and gel-forming characteristics, these possibilities were not pursued. However, it was noted that compared to native casein, the reductively formed sugar derivatives of casein had lower in vitro digestibility by α-chymotrypsin and lower nutritive values in rat-feeding experiments.

D. Lipophilization

In an attempt to enhance the amphipathic nature of food proteins, modification by "lipophilization" has been suggested. According to Aoki et al.,[69] the term lipophilization refers generally to the increase in hydrophobicity of proteins by modification, with a consequent increase of affinity for relatively nonpolar compounds. In this sense, the incorporation of hydrophobic groups by esterification, amidation, acylation or reductive alkylation may all be referred to as "lipophilization". In this section, however, the term has been used to refer

Table 5
FUNCTIONAL PROPERTIES OF HIGHLY ALKYLATED CASEINS COMPARED TO NATIVE CASEIN

Casein sample	Relative viscosity[a]	Absorbed moisture,[b] (%)	Relative emulsifying activity[c]
Native	1.0589	0.48	1.00
Methyl	1.0171	0.71	1.64
Isopropyl	1.0367	0.84	1.81
Butyl	1.0496	0.40	1.00
Cyclopentyl	—	0.39	1.48
Cyclohexyl	1.0525	0.97	1.33
Benzyl	—	0.67	1.53

[a]   Viscosity at 25°C of 0.1% protein solution, relative to dissolving buffer (0.02 M borate, pH 8.2).
[b]   Measured in 18.8% relative humidity, 25°C.
[c]   Emulsifying activity of 0.40% protein solution in 0.1 M phosphate buffer pH 7.0 at 25°C, measured at protein solution: oil volume ratio of 10:3, and expressed on a relative scale to native casein.

Adapted from Sen et al.[67]

primarily to those types of modification which specifically involve attachment of fatty acids, fatty acid derivatives, or surfactant molecules to proteins, or to the interaction of alcohols with proteins.

Long chain fatty acids covalently attached to proteins do occur naturally from posttranslational in vivo chemical modification, and have been identified in a number of membrane-associated proteins. For example, palmityl, palmitoleyl, and *cis*-vaccenal groups linked to the protein amino termini were identified in bacterial outer-membrane proteins,[70] while myristic acid in amide linkage to amino termini were identified in murine leukemia virus membrane protein,[71] chicken embryo fibroblast membrane proteins, and cyclic AMP-dependent protein kinase from bovine cardiac muscle.[72] The covalently bound fatty acyl groups were suggested to contribute greatly to affinity for hydrophobic surfaces including cellular and viral membranes, and in vitro to reverse phase-HPLC supports.[71] The lipoprotein on the outer membrane of *E. coli* shows a regular distribution pattern of hydrophobic amino acids in the amino acid sequence, with hydrophobic amino acids at every 3.5th position.[70] Since 3.6 residues make up a helical turn, all of the hydrophobic residues would be aligned on one face of the helical rod, as supported by circular dichroism spectral data. This unique pattern of amino acid residues in the lipoprotein has been speculated to be responsible for aggregation by hydrophobic interactions, interactions with fatty acids, and lipophilic membrane components, and for the unusually high degree of ordered structure of the lipoprotein in an aqueous environment. Measurements of the circular dichroism of the lipoprotein suspensions revealed an α-helical content of about 80%, and this ordered helical structure was completely regained after renaturation of heated- or urea-denatured lipoprotein. These unique characteristics of naturally occurring proteins with covalently attached fatty acyl groups should prompt investigations of properties of proteins deliberately modified by lipophilization.

Lipophilization of a hydrophilic enzyme, α-chymotrypsin, was achieved by Torchilin et al.[73] via acylation of the protein amino groups with palmitic chloroanhydride. Between one

Table 6
HLB NUMBER, PROTEIN SOLUBILITY, AND
EMULSION STABILITY OF SOY PROTEIN AND
MODIFIED DERIVATIVES

Soy sample	HLB	Protein solubility,(%)	Emulsion stability, (%)
Unmodified	10.3	5	5
Acetylated			
30%	10.3	5	15
70%	9.7	5	25
95%	8.9	5	45
n-Propanol modified	9.3	5	59
Partially hydrolyzed			
Whole fraction	9.3	63	80
Soluble fraction[a]	9.7	100	7
Insoluble fraction[a]	8.9	0	83

[a]   Soluble and insoluble fractions were the supernatant and pellet
      fractions, respectively, after centrifugation of pH 4.5 solution at
      8000 rpm for 20 min.

Adapted from Aoki et al.[69]

to six palmitoyl fatty acid residues were incorporated per enzyme molecule, with little change in the modified enzyme's catalytic properties. The palmitoyl chymotrypsin could be effectively incorporated both actively and passively into model liposomal membranes. Electron spin resonance and fluorescence-spectroscopic studies indicated that the hydrophobic tail of the modified enzyme was incorporated into the membrane while the protein globule was located on the surface of the membrane.

The effects of lipophilization of soy protein on emulsion-stabilizing properties were investigated by Aoki et al.[69] Modification of soy protein by incubation with 50% alcohol (ethanol, iso-propanol, and *n*-propanol) at 35°C for 2 hr increased emulsion stability in the acidic pH range, particularly at the isoelectric point. Similar changes in the pH-emulsion stability profile were observed upon modification by acetylation with acetic anhydride or by partial hydrolysis with dilute hydrochloric acid. Aoki et al.[69] speculated that lipophilization had a close connection with soy protein emulsifying properties. Alcohol-modified, acetylated, and partially hydrolyzed soy derivatives generally had higher hydrophobicity than the unmodified soy protein, as indicated by their lower HLB numbers (Table 6), higher relative fluorescent intensity upon ANS binding (Table 7) and increased exposure of tyrosine residues measured from ultraviolet difference spectra.[69] The extent of alcohol modification could be altered by changing the concentration of alcohol, and the temperature and time of incubation (Table 7). The resulting differences in hydrophobicity as measured by relative fluorescence intensity appeared to be related to both intrinsic viscosity and pH 4.5 emulsion stability. However, the emulsifying properties were suggested to ultimately depend on suitable balance between hydrophile and lipophile. Excessive lipophilization or denaturation would be expected to result in lower emulsion-stabilizing properties. Under mild conditions of alcohol modification (23°C for 2 hr), the emulsion stability at pH 4.5 increased in the following order: ethanol < iso-propanol < *n*-propanol. However, under more severe conditions (50°C for 24 hr), the order of emulsion stability was reversed: *n*-propanol < iso-propanol < ethanol, which may be related to the denaturing capacity of the alcohols.

The hydrophilic protein soybean glycinin was lipophilized by base-catalyzed ester exchange using the *N*-hydroxysuccinimide esters of naturally occurring fatty acids of various

Table 7
**EFFECTS OF DIFFERENT CONDITIONS OF
ETHANOL MODIFICATION OF SOY PROTEIN ON
HYDROPHOBICITY (ANS RELATIVE
FLUORESCENCE INTENSITY), INTRINSIC
VISCOSITY, AND EMULSION STABILITY**

Soy sample	ANS relative fluorescent intensity	Intrinsic viscosity (dl/g)	pH 4.5 emulsion stablity, (%)
Unmodified	43	0.061	5
Ethanol-modified (A)[a]	182	0.086	37
Ethanol-modified (B)[a]	214	0.242	44
Ethanol-modified (C)[a]	311	0.280	61

[a] Ethanol modification conditions were as follows: (A) 10% ethanol, 20°C, 5 min; (B) 50% ethanol, 20°C, 2 hr; (C) 50% ethanol, 35°C, 2 hr.

Adapted from Aoki et al.[69]

lengths, including lauric, myristic, palmitic, and oleic acids.[74] Since the association of phosphatidylcholine with soybean proteins was reported to be due to hydrophobic interactions, Kito and co-workers[74,75] decided to investigate the effects on soybean protein functional properties by covalent anchoring of hydrophobic groups. In the native soybean glycinin, the hydrophobic core is tucked away from the aqueous environment. It was anticipated that if sufficient hydrophobic sites could be created on the protein surface without adversely affecting the solubility, the amphiphilic nature and functional properties of the lipophilized protein would be enhanced. In fact, covalent attachment of palmitoyl residues to soybean glycinin resulted in improvement in emulsification activity, foam activity, and foam stability.[75] Interestingly, the palmitoyl proteins remained soluble in spite of their increased hydrophobicity.

Subsequently, palmitoyl α_{s1}-casein derivatives were prepared by Haque, Kito, and co-workers, both by the *N*-hydroxysuccinimide ester method[76-78] and by a one-step method using fatty acyl anhydrides.[79] Unlike the hydrophilic acidic soybean glycinin protein, α_{s1}-casein is itself hydrophobic, about 40% of its amino acids containing hydrophobic side chains, many of which could be exposed. Logically, lipophilization by attachment of hydrophobic ligands to α_{s1}-casein would be expected to have little effect or even adverse effect on its amphipathicity. However, it was speculated that covalent attachment of long chain fatty acids would create not only new hydrophobic sites, but by virtue of the greater spatial flexibility of the tail-like hydrophobic ligand, would perhaps geometrically favor greater association by hydrophilic-lipophilic arrangement. The palmitoyl α_{s1}-casein did indeed show an enhanced tendency to associate in micellar aggregates and exhibited "soap-like" behavior, especially when six ligand molecules were attached.[76] Tendency to associate decreased at higher levels of palmitoyl incorporation, probably due to imbalance of amphipathic forces or due to increased net negative charges upon linking of palmitoyl residues to lysine amino residues.

The amphipathic nature of α_{s1}-casein was improved dramatically upon covalent incorporation of palmitoyl residues, as manifested by an improved ability to form and stabilize emulsions, and by the increase in foam stability and foam activity by incorporation of up to six palmitoyl residues per mol of protein.[77] The improvement in these properties was attributed primarily to the incorporation of the hydrophobic palmitoyl substituents per se, since no drastic conformational changes were observed by circular dichroism studies.

One-step lipophilization of soybean flakes and casein using fatty acyl anhydrides indicated that as the length of the fatty acyl ligand increased (caproic < capric < myristic), surface tension of the lipophilized protein decreased.[79] Minimum surface tension was obtained at a level of incorporation of two myristoyl residues per casein molecule, with higher levels of incorporation probably having an adverse effect on surface amphipathicity.

Novel amphipathic fatty acyl peptides were obtained by enzymatic hydrolysis of lipophilized α_{s1}-casein, containing covalently attached caprylic, lauric, myristic, palmitic, stearic, oleic, or linoleic acid.[78] These amphipathic fatty acyl peptides showed a tendency to form micelle-like aggregates in units of 3, 6, 12, and 24, with the trimer being the most predominant. The driving force for initial aggregation to form the trimer was spontaneous, but further aggregation seemed to depend on ligand size, with longer ligands leading to larger aggregates. Critical micelle concentration also depended on both ligand length and degree of unsaturation. Surface tension decreased and foaming activity was higher when ligand length was longer, while foam density decreased with increasing ligand length and unsaturation.

Effects of lipophilization have also been investigated by noncovalent attachment or binding of fatty acids or their salts, which are also referred to as detergents, surfactants, or soaps. The insolubility in water of the wheat protein glutenin was improved markedly by incorporation of sodium salts of long chain fatty acids such as sodium palmitate or sodium stearate.[80] Sodium salts of shorter chain fatty acids were less effective for solubilization, while except for hexanoic acid, fatty acids were ineffective for solubilization. The ability of soaps to dissolve glutenin was attributed to the interaction between their hydrophobic chains and the hydrophobic regions of glutenin.

Small amounts of anionic detergents were shown to have a protective effect on the thermal aggregation of crude ovalbumin.[81] Low concentrations (1 to 5 mM) of 2-decylcitric acid, sodium dodecylsulfate, or lauric acid were able to prevent precipitation of ovalbumin even when the temperature was increased to 100°C, while no comparable raise in aggregation temperature was observed if a nonionic detergent (n-dodecanol) was used. It was proposed that the penetration of the hydrocarbon chain of the detergent molecule into the internal regions of the protein caused internal stabilization of the more hydrophobic center toward thermal stress, whereas the anionic group could increase repelling Coulomb's forces. It was suggested that this type of modification could have the potential for extending use of normally coagulating types of food proteins in products where no protein aggregation is desired.

Similar effects were noted for fatty acids and their salts on stability of serum albumin against denaturation by heat or urea and guanidine hydrochloride.[82,83] Low concentrations of the fatty acid salts had a pronounced protective effect against serum albumin denaturation, with the effect being more pronounced with increasing chain length of the fatty acid. The effect was relatively independent of ionic strength and pH in the neutral and acidic range, and both polar and nonpolar interactions between the fatty acid salts and protein were suggested to be responsible for the stability. Heat treatment of serum albumin itself was shown to cause it to behave as a heterogeneous population of molecules.[84] Prolonged heating at 60°C resulted in formation of an aggregate fraction and a fraction that was stable against further heat treatment. Fatty acid analyses indicated the heat-stable fraction was formed by migration of fatty acid molecules from aggregating molecules to the remaining monomers, thereby stabilizing the latter against heat denaturation.

Solubilization of rapeseed, soy, and sunflower protein isolates by anionic surfactant treatment was compared to solubilization by proteinase treatment.[85] Sodium and potassium salts of myristic, oleic, and linoleic acids and sodium dodecylsulfate were all effective for improving dispersibility or solubilization, as were enzymatic proteolysis using *B. subtilisin* protease, trypsin, or pepsin.

The amount of bound dodecylsulfate or linoleate was significantly positively correlated to dispersibility of the treated proteins.[86] Dispersibility was also highly correlated with the effective hydrophobicity determined with a CPA fluorescence probe. Although this appears to be contrary to the general concept of hydrophobicity-solubility relationships, it is possible that an increase in electrostatic repulsive forces due to the negative charges of the anionic surfactants, may have superseded the effect of increased hydrophobicity on solubility. Increase in hydrophobicity of the surfactant-treated proteins was attributed to greater exposure of hydrophobic sites on treated proteins as well as the possibility of an increase in hydrophobic sites due to the hydrophobic nature of the surfactants themselves. The importance of the increased hydrophobicity of surfactant-treated proteins was also demonstrated from their significant correlations with surface tension, interfacial tension, and emulsifying activity. Solubilization by surfactant treatment was preferable to proteinase treatment for soy, sunflower, and rapeseed proteins due to the additional benefit of emulsifying property improvement.

Linoleate treatment of oilseed protein concentrates and isolates was demonstrated generally to improve not only solubility but also water-hydration capacities.[87] Three possible factors were suggested to contribute to this improvement of water-hydration capacity—disruption of protein-protein hydrophobic groups, increase in net negative charge from binding of the anionic ligand, and the increase in mean particle size of linoleate treated proteins in dispersion.

Functionality of trypsin, potassium linoleate and sodium dodecylsulfate-solubilized plant proteins was tested in model systems of whipped topping, oil-in-water emulsion and model frankfurter emulsion.[88] Although enzyme and surfactant treatments all increased solubility of the plant proteins, solubility was a significant factor only for foam volume of the whipped topping model system. For emulsion viscosity and stability and for frankfurter cook stability and firmness, factors related to protein-lipid interactions were most significant, including apparent hydrophobicity, emulsification capacity, emulsifying-activity index, and fat absorption.

These studies on lipophilization point out the potential for extending applications of food proteins, including the ability to form micellar-like aggregates,[78] the incorporation of proteins into liposomes,[73] and even the possibility of forming reverse micelles as a bioseparation tool,[89] by enhancing the amphiphilic nature of the lipophilized proteins and improving their functional properties.

E. Glycosylation

In nature, glycosylations are difficult to assign to a single functional classification, as it appears that many different biological functions can be ascribed to sugar moieties; in some cases, they may establish structural integrity, protease resistance, increased solubility, or high viscosity, while in other cases, they may fulfill more dynamic functions, including specific biological recognition.[90] In many food systems, naturally occurring glycoproteins impart excellent emulsion and foam stability characteristics.[91] These important functional properties of glycoproteins reflect chemical and structural characteristics of the carbohydrate moieties as well as the proteins.

Glycosylation of proteins by chemical modification has been suggested in order to confer on the glycosylated proteins some of the desirable functional properties which are evident in natural glycoproteins. Covalent attachment of glycosyl moieties has been achieved using cyanogen bromide,[92,93] glutaraldehyde,[92] water-soluble carbodiimide,[94,95] cyclic carbonates,[95] and sodium cyanoborohydride.[68,96]

Increased stability of enzymes against heat, denaturing agents, and proteolysis has been reported by dextran conjugation of β-glucosidase,[92] α- and β-amylases,[93] lysozyme,[92] and trypsin.[93] Similarly, the carbohydrate moieties of naturally occurring glycoproteins such as glucose oxidase, have been reported to enhance the protection towards denaturation by sodium dodecylsulfate.[97] The mechanisms of enhanced stability by glycosylation are not clear. Marshall and Rabinowitz[93] suggested increased degree of hydration, cross-linking, and prevention of unfolding of the glycosylated molecules as possible explanations. On theother hand, Back et al.[98] suggested that the stabilization of proteins against heat denaturation in the presence of sugars and polyols could be explained by the effect on hydrophobic interactions. Sucrose and glycerol strengthened pairwise hydrophobic interaction between hydrophobic groups and they reduced the tendency for complete transfer of hydrophobic groups from an aqueous to a nonpolar environment. It was postulated that the enhanced hydrophobic interactions stabilized the protein structure against denaturation.

In addition to the protective effect of glycosylation against denaturation, other effects such as improved solubility have been observed. For example, covalent attachment of gluconic or melibionic acids to β-lactoglobulin resulted not only in improved heat stability, but also in high solubility of the glycosylated protein, even at low ionic strength or at the isoelectric point of native β-lactoglobulin.[94] These improvements in functional properties increased with the extent of modification, with the disaccharide melibionic acid being more effective on a molar basis than the monosaccharide, gluconic acid. Covalent binding of a small number of glycosyl residues was suggested to be specifically responsible to the stabilization, since control β-lactoglobulin in the presence of free gluconate or melibionate showed no heat stabilization effect. The effects of glycosylation were speculated to be due both to changes in the net charge of the protein and to the introduction of the hydrophilic residues, but structural properties of the glycosylated protein were not investigated.

Maltosyl, beta-cyclodextrinyl, glucosaminyl, and glucosamineoctaosyl derivatives of β-lactoglobulin were prepared to determine the effect of added carbohydrates of different sizes on the structural and functional properties.[95] Oligosaccharide (i.e., beta-cyclodextrinyl and glucosamineoctaosyl) derivatives had an increased coupling efficiency which was attributed to strong noncovalent interactions of the oligosaccharide moieties with the proteins prior to the coupling reaction. Electrophoretic mobility of maltosyl β-lactoglobulin derivatives was essentially unchanged while mobility of glucosaminyl β-lactoglobulin derivatives decreased as the extent of modification increased. Decreased mobility was thought to arise from the added mass of the carbohydrate residues, while in the case of maltosyl derivatives, increased charge-to-size ratio was suggested to compensate for the increased size by maltosyl attachment.

Physicochemical properties of glycosylated β-lactoglobulin derivatives were analyzed.[100] Viscosity increased with increasing mass of the covalently attached carbohydrate. Differences in protein structural properties were noted between maltosyl and glucosaminyl derivatives of β-lactoglobulin. Whereas the maltosyl derivatives underwent conformational changes reflected by increasing changes in properties with increasing extent of modification, the glucosaminyl derivatives seemed to show significant large changes in tertiary conformation only at high levels of modification (>16 glycosylated residues). Ultraviolet difference spectra and intrinsic fluorescence revealed that glycosylation with maltosyl residues resulted in increasing exposure of aromatic amino acid residues to the surface of the protein molecules, resulting in increased polarity of the microenvironment of these aromatic residues with increasing modification. On the other hand, aromatic amino acid residues of glucosaminyl derivatives were in a less polar environment and intrinsic fluorescence of the glucosaminyl derivatives were either unchanged or slightly higher than unmodified β-lactoglobulin.

Hydrophobicities determined using a *cis*-parinaric acid fluorescence probe were decreased for maltosyl derivatives, while hydrophobicities of glucosaminyl derivatives were higher than unmodified β-lactoglobulin at levels of modification of up to 16 glucosaminyl residues, then decreased with greater extent of modification. The changes in *cis*-parinaric acid fluorescence intensity would reflect not only the number and nonpolarity of hydrophobic-binding sites on the protein molecules, but also possible steric interference of probe binding by the hydrophilic carbohydrate moieties.

The conformation of the protein was also affected by type and extent of glycosylation, as determined by circular dichroism. Maltosyl derivatives showed decreasing α-helical structure and increasing unordered, random coil structure with increasing extent of modification. On the other hand, glucosaminyl derivatives showed increased in β-structure at intermediate levels of modification up to 16 glucosaminyl residues; decrease in α-helical structure and increase in unordered structure were only observed at higher levels of modification of about 22 glycosylated residues.

Foaming properties of the glycosylated β-lactoglobulin derivatives were improved compared to the unmodified protein.[99,101] The amount of drainage of liquid from foams prepared with glycosylated derivatives were generally less than that of foams prepared with unmodified protein. Foam strength was also improved, including at pH 5, which is near the isoelectric point of unmodified β-lactoglobulin. Foam strength of β-lactoglobulin modified with an oligosaccharide derivative of glucosamine (glucosamineoctaosyl) was, however, only slightly higher than unmodified protein.

In vitro digestibilities using trypsin and α-chymotrypsin indicated an increasing rate of hydrolysis with an increasing number of attached carbohydrate residues, especially for α-chymotrypsin digestion.[102] Maltosyl derivatives were digested at a faster rate than glucosaminyl derivatives, attributed to their greater extent of unfolding and more expanded tertiary conformation. It was suggested that the extensively glycosylated derivatives had larger numbers of basic and aromatic amino acids exposed at the protein surface, which resulted in the higher rates of hydrolysis. Less extensively modified derivatives, on the other hand, showed lower rates of hydrolysis, due possibly to steric hindrance by the attached carbohydrate residues in conjunction with little change in exposure of amino acids for derivatives which had been glycosylated to lower levels.

F. Phosphorylation

Phosphoproteins occurring naturally in food systems include milk caseins, egg white ovalbumin, and egg yolk phosvitin. These phosphoproteins are formed in vivo by phosphorylation of the protein precursors via a reaction catalyzed by the enzyme phosphoprotein kinase. Phosphorylation of proteins can be achieved in vitro by a variety of chemical covalent attachment techniques, as reviewed by Kester and Richardson,[17] and Matheis et al.[103] The most common reaction for chemical modification with phosphate groups is by reaction with phosphorus oxychloride. Alternative phosphorylating reagents include phosphorus pentoxide dissolved in phosphoric acid, cyclic sodium trimetaphosphate, phosphoramidate, phosphoric acid combined with trichloroacetonitrile as a coupling agent, monophenyl phosphodichloride, and diphosphoimidazole. These chemical derivatizations may attach the phosphorus-containing moiety to the hydroxyl oxygen of serine and threonine residues, the amino nitrogen of lysine residues or the imidazole nitrogen of histidine residues.

Because protein surfaces are generally already quite hydrophilic, the attachment of phosphate groups may not exert as dramatic an effect on functionality as attachment of hydrophobic groups.[17] Nevertheless, changes in functional properties have been observed in food proteins upon phosphorylation,[103-106] including changes in solubility, water absorption, viscosity, gel-forming and surfactant (emulsifying) properties. Despite these observations of changes in functionality, however, investigations into the corresponding structural and physicochemical changes in the phosphorylated proteins have been scarce. It is anticipated that phosphorylation for improvement of properties may be better controlled if systematic investigations on the structural changes are initiated.

Phosphorylated casein containing mono- and diphosphates bound to hydroxyl oxygen of amino acid residues had significantly higher relative viscosity and water-binding properties than control casein, whereas emulsifying activity was significantly lower after phosphorylation.[103] Phosphorylated lysozymes contained mono-, di-, and triphosphates bound to nitrogen of amino acid residues and showed greatly improved water-binding properties compared to control lysozyme, but did not differ in relative viscosity. Both phosphorylated casein and lysozyme showed lower solubility than the unmodified proteins, which may have been due to formation of protein crosslinks by phosphorylation.[90]

Soy protein isolate modified using cyclic sodium trimetaphosphate exhibited much-improved functional properties in terms of aqueous solubility, water-holding capacity, emulsifiability and whippability.[104] It was suggested that the covalent attachment of anionic phosphate groups to polypeptide chains and the increase in net electronegativity altered the physicochemical character of soy protein, resulting in enhanced functionality. These changes were likened to the effects of succinylation, in which the modification alters not only the net charge, but also the conformation and exposure of polar and hydrophobic regions to the protein surface.[42,46] However, the alterations in physicochemical and structural properties have not been similarly investigated in the case of phosphorylation, so the mechanisms for improvement in functionality are only speculative at present.

Structural changes and functional properties of phosphorylated bovine β-lactoglobulin were studied by Woo and co-workers.[105,106] Irreversible changes were suggested to result from phosphorylation, since indistinguishable spectra between dephosphorylated phosphoprotein and the phosphorylated protein were observed; these spectra were different from the unmodified protein. It was suggested that covalent bonds such as isopeptide linkage may have been formed by phosphorus oxychloride treatment, which would irreversibly constrain the molecules in a conformation different from the unmodified protein. Partial unfolding as well as some dimerization of protein molecules were suggested to occur upon phosphorylation.

Corn oil emulsions (65% w/w) prepared with phosphorylated β-lactoglobulin exhibited almost twice the relative viscosity compared to emulsions prepared using unmodified β-lactoglobulin. Emulsion stability was improved and sensitivity towards calcium ions were altered, using phosphorylated protein. Of particular interest was the ability of the phosphorylated protein to gel in the presence of 100 mM Ca^{2+} at pH 5.0 and room temperature. These improved emulsifying and gel-forming properties were suggested to be useful for the application of phosphorylated whey proteins in food systems such as dressings and mayonnaise.

G. Arylsulfonation and Alkylsulfonation

Reaction of sulfonates and sulfonyl chlorides with proteins can be used to introduce hydrophobic alkyl and aryl sulfonyl moieties. Sulfonates react primarily with the amino groups on amino acid residues, while the sulfonyl chlorides may react with hydroxyl and sulfhydryl groups as well as amino groups.[107]

The effects of aryl- and alkyl-sulfonating reagents on the setting of muscle proteins and muscle flesh sols of different species were studied by Niwa et al.[107-109] The phenomenon of "setting" observed in some fish species, wherein an elastic gel is formed on standing or mild heating of the salted fish mince, was hypothesized to arise from hydrophobic interactions of protein amino acid residues. The inability to set is observed in species such as chicken, beef, and pork and also in some fish species. It was speculated that increasing the hydrophobicity of the surface of proteins from these hard-to-set species by introduction of aryl and alkyl groups by chemical modification, would induce "artificial" setting. Their results showed that ethylsulfonation as well as arylsulfonation both induced proteins and sols of nonsetting species to behave like easily-setting species, exhibiting characteristics of increased water binding, increased viscosity upon heating and induction of gelling at low temperatures of 4 and 40°C. These observations suggested that both aliphatic and aromatic groups are related to the setting phenomenon and that hydrophobic interactions are crucial in this phenomenon. In the chemically modified proteins, the hydrophobic interactions would be formed easily since the hydrophobic groups would be introduced preferentially to the protein surface. However, in the case of naturally easily-setting species, it is unknown whether only the hydrophobic residues existing originally at the molecular surface are responsible for setting, or whether residues buried in the interior may become exposed during setting.[108]

H. Chlorination

Freshly milled wheat flour has a light yellow color and yields a sticky dough with poor handling and baking properties. Upon storage, the flour undergoes an aging or maturing process which improves its baking qualities; however, to accelerate these natural processes, chemical treatments are often applied. Oxidizing agents commonly used as bleaching and improving agents for wheat flour include chlorine gas, chlorine dioxide, nitrosyl chloride, and oxides of nitrogen.[110] These oxidizing agents are gaseous and exert their action immediately upon contact with flour. Their influence on sulfhydryl and disulfide groups of flour proteins and on the characteristics of flour lipids, are generally accepted as the basis behind improvement in dough properties.

Observations of strong oil-binding ability in the chlorinated starch of wheat, corn, potato, rice, and arrowroot, and the decrease of this oil-binding ability by protease digestion indicated that chlorination may also have the effect of increasing the hydrophobicity of the thin protein film on the starch granule.[111-113] In order to investigate the role of protein in the lipophilization, model experiments on hydrophobicity of chlorinated starch and chlorinated surface protein were carried out.[114] Chlorination of gelatin-coated glass powder or glass beads imparted strong oil-binding ability similar to that exhibited by chlorinated starch. Chlorination of films of proteins on petri dishes rendered the normally soluble proteins (gelatin, bovine serum albumin, and ovalbumin) into hydrophobic, water-insoluble films. On the other hand, no film formation or increase in hydrophobicity was observed by chlorination of soluble starch, indicating that the change in hydrophobicity upon chlorination was specific to protein and not to starch. The water-insoluble washed protein films could be readily dissolved in weak alkali or in 1% sodium dodecylsulfate (SDS) solution, and could be reprecipitated to form protein flocculants by dialysis against water. SDS disc gel electrophoresis showed no differences in molecular weight of nonchlorinated and chlorinated protein. An increase in the ultraviolet absorption between about 260 to 300 nm for chlorinated protein derivatives suggested the role of aromatic amino acids such as tyrosine in the increase in hydrophobicity by chlorination. Paper chromatography showed almost no change in any of the amino acids, except tyrosine, lysine, and cystine; for the latter three amino acids, R_f of chlorinated derivatives were higher than that of nonchlorinated ones, suggesting a change to a more hydrophobic nature. These results were suggested to support the hypothesis that lipophilic properties acquired by chlorination of wheat starch may be attributed to a lipophilic protein film on the starch granule; lipophilization of chlorinated protein film may be induced by hydrophobic molecules of chlorinated amino acids such as tyrosine, lysine, or cystine.

I. Amino Acid Attachment

Amino acids have been successfully attached covalently to food proteins by chemical modification, using *N*-carboxy-α-amino acid anhydrides,[115] *N*-hydroxysuccinimide esters of acylated amino acids,[116-118] and by water-soluble carbodiimide-mediated coupling of amino acids.[119-123] These modifications were performed for the purpose of nutritional improvement of food proteins by covalent attachment of essential amino acids through peptide or isopeptide linkage. Only limited information is available on changes in structural and functional properties of these nutritionally improved proteins.[118] However, the potential exists for changing protein-hydrophobic interactions by attachment of hydrophilic or hydrophobic amino acids, including amino acid polymers or oligopeptides. Improvement in biological properties have been reported for albumin and asparaginase by conjugation with poly-L-lysine and poly-D,L-alanyl peptides, respectively.[118] Improvement in functional properties of proteins after amino acid attachment by an enzymatically catalyzed reaction has also been reported, as described in the next section on enzymatic modification.

III. ENZYMATIC MODIFICATION

The application of enzymatic modification procedures to food proteins has been the subject of several symposia in recent years.[124-129] Endogenous enzymes are responsible for hydrolytic reactions which may have great significance in the quality of foods, such as tenderization of meat, development of ripened cheese characteristics, development of off-flavors, and spoilage of foods. In addition, in vivo, enzymes catalyze many posttranslational modifications of proteins, including crosslinking of polypeptide chains, phosphorylation, glycosylation, hydroxylation, and methylation.[128] In this section, the reactions dealing with enzymatic derivatization of amino acid side chains such as phosphorylation and glycosylation will not be covered. Although the specificity of enzyme-catalyzed reactions may influence the modification, the properties of the enzymatically modified derivatives would probably be generally similar to those prepared by chemical modification methodologies, as described in the previous section. This section, therefore, mainly deals with the enzymatic modification of food proteins by proteolytic enzymes, including the hydrolysis and resynthesis of peptide bonds.

A. Enzymatic Hydrolysis

Proteolytic enzymes have been widely used for partial hydrolysis to improve functionality of a variety of food proteins, including proteins from soy,[130-133] rapeseed,[85-88,134] sunflower,[85,86] fish,[135-137] beef skeletal[138] and heart muscle,[139] mechanically deboned fowl,[140] whey,[141] wheat,[142,143] and egg.[144] The hydrolysis of peptide bonds can result in an increase in charged groups (extra amino and carboxyl termini) and hydrophilicity, decrease in molecular weight, and alterations in molecular configuration.[125] Thus, the peptide bond breakage may be followed by a complex series of changes which can alter functionality through dissociation of subunits or unfolding of a compact structure to expose hydrophobic regions. Commonly, an increase in solubility and a decrease in viscosity are observed with increasing hydrolysis. Other effects which are frequently observed include altered gelation properties, enhanced thermal stability, increased emulsifying and foaming abilities, and decreased emulsion and foam stabilities.

The specific properties of each hydrolyzed product depend on the degree of hydrolysis, the specific enzyme used, the substrate, and hydrolysis conditions. Unfortunately, there are few studies which have been specifically designed to quantitatively relate the parameters characterizing protein hydrolysates to their functionality.[126,131] The need for basic research in this area was stressed by Richardson.[126] Until such studies are conducted, the effects of enzymatic hydrolysis on functionality cannot be fully understood or predicted. For example, Puski[145] reported that soy protein hydrolysate prepared by proteolysis with *Aspergillus oryzae* neutral protease, significantly reduced viscosity of concentrated protein solutions and prevented their gel formation. On the other hand, Pour-El and Swenson[146] were able to form heat-induced gels from various soy protein hydrolysates prepared using ficin, bromelain, papain, pepsin, trypsin, pancreatin, and acid fungal protease. Products with varying gel characteristics were obtained, depending on the kind of enzyme used for hydrolysis, with plant protease treatment generally producing better gels than animal enzyme treatment. Fuke et al.[147] showed that heating of the water-soluble soy protein extracts resulted in formation of soluble aggregates; subsequently, enzymatic treatment of the heat-induced soluble aggregates using bromelain resulted in formation of insoluble aggregates and a self-supporting gel. It was suggested that bromelain treatment caused partial unfolding of the soluble aggregates, leading to aggregation and gelation. Based on measurement of protein surface hydrophobicity using ANS, noncovalent forces, particularly hydrophobic interactions, were postulated to play an important role in the aggregation process.

An important consideration in the use of proteolysis to improve functionality is the formation of bitter peptides upon hydrolysis of certain proteins, including casein and soy.

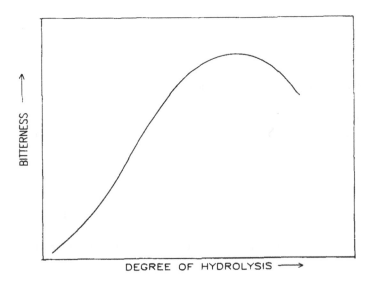

FIGURE 1. Proposed relationship between bitterness and degree of hydrolysis of proteins (Modified from Alder-Nissen and Olsen.[130])

Ney[148,149] proposed that bitter peptides have an average hydrophobicity of at least 1400 cal res^{-1}, and that the amino acid sequence of the peptides is not related to bitterness; proteins with high-average hydrophobicity were suggested to have a tendency to yield bitter peptides upon hydrolysis. Although bitterness is generally related to hydrophobicity of the amino acids in the peptides, this effect depends on the extent of exposure of the hydrophobic side chains to the solvent to allow contact with taste-bud receptor sites. Small peptides existing in solution as random coils would thus have more exposed hydrophobic residues than large peptides or polypeptides which form more compact secondary and tertiary structures. In addition, hydrophobic amino acids whose amino and carboxyl groups are blocked, for example through peptide bond formation, are more bitter than the free amino acids.

Adler-Nissen[130] and Adler-Nissen and Olsen[131] suggested that the number-average peptide chain length is inversely proportional to the degree of hydrolysis (DH), which is a principal determinant of taste and functionality of the hydrolysates. The qualitative relationship between bitterness and DH of protein hydrolysates suggested by Adler-Nissen and Olsen[131] is depicted in Figure 1. The following model was proposed by them to explain the relationship between bitterness and DH.[131] In the intact protein or at low DH values, the majority of hydrophobic side chains are masked and unavailable to taste-bud receptors, thus bitterness is low. At increasing DH values, the size of the peptides formed may become too small to form proper secondary structure, thus hydrophobic groups are increasingly exposed and bitterness is increased. At very high levels of DH, an increasing proportion of the hydrophobic amino acids are either in a terminal position or exist as the free form, thus reducing the bitter taste.

Methods for debittering protein hydrolysates have been suggested. Lalasidis and Sjoberg[150] reported that removal of 5 to 10% of the hydrolysate by extraction with azeotropic secondary butyl alcohol was effective for complete debittering of protein hydrolysates from soy, fish protein concentrate, deboned cod filleting offal, and fresh herring. In particular, tryptophan, phenylalanine, leucine, tyrosine, valine, and isoleucine were removed by the alcohol extraction. Plastein formation was also suggested for debittering these hydrolyzates. Alternatively, a reduction in bitterness could be achieved using hydrophobic interaction chromatography on hexyl Sepharose®, although the gel had only a limited capacity for binding the bitter compounds.

Bitterness of casein hydrolysates could not be reduced by plastein reaction,[151] but could be almost completely eliminated by hydrophobic chromatography on hexylepoxy Sepharose®.[152] Activated carbon, talc, and β-cyclodextrin had higher debittering capacities for casein hydrolysates than Sephadex® G-10 or G-25.[152] Chromatography on activated carbon, glass powder, or fiber, Sephadex® LH-20 and phenoxyacetyl cellulose were effective, in that order, for reducing bitterness of skim milk hydrolysates.[152]

B. Plastein Formation

Proteolytic enzymes are generally associated with catalysis of peptide bond hydrolysis, giving rise to products with lower molecular weight and higher water solubility than the original substrate. However, if highly concentrated solutions of protein or peptide hydrolysates are incubated with proteases under appropriate conditions, water-insoluble and/or gel-forming products may be formed. The term "plastein" was first used by Sawjalow[153] to describe these resynthesis products.

In order for the plastein reaction to proceed efficiently, the reaction conditions must be rigidly controlled. The rate and extent of protein degradation which yields the starting material for the plastein synthesis are crucial. Factors which affect the degradation or hydrolysis step include specificity of the enzyme, extent of denaturation of the protein substrate, concentrations of enzyme and substrate, pH, temperature, and ionic strength of the reaction medium.[124] Following the degradation to yield the peptides, the enzymatic resynthesis to form the plastein product must also be carried out under specific conditions. Among the most important parameters to consider are substrate concentration, average molecular weight of the substrate, and pH of the reacting medium.[124,154,155] Substrate concentration should be in the range of 20 to 40% (w/v),[124] since at lower substrate concentrations, the tendency is for hydrolysis rather than synthesis.[156] The substrate must be of low molecular weight and it has been reported that a peptide size of four to six amino acid units is most favorable.[157,158] The addition of an organic solvent to the system usually favors plastein formation. The optimal pH generally lies in the range of 4 to 7 for resynthesis, in contrast to the optimum pH for hydrolysis which may vary from 1.6 for hog pepsin to 10 to 11 for Bioprase, a serine protease from *B. subtilis*.[159] Based on the difference between the optimum pH values for synthesis and hydrolysis (ΔpH), Yamashita et al.[159] classified proteases into three types: pepsin type (Δ pH >0), chymotrypsin type (Δ pH <0), and papain type (Δ pH =0), with the exception of trypsin (not plastein-productive at any pH). Among the well-known proteases, comparatively high plastein yields were obtained using pepsin, α-chymotrypsin, and papain.

Hydrophobic interactions in the plastein reaction are implicated by the favorable effect of solvent on the reaction, the water insolubility of the product, and the rates of incorporation of different amino acids into plastein product. From model experiments using *N*-benzoyl-glycine ethyl ester or ethyl hippurate as the substrate, instead of protein hydrolysate, Aso et al.[160] noted that various L-amino acid ethyl esters were incorporated via papain-catalyzed peptide bond formation at different rates, with amino acids carrying hydrophobic side chains being more effectively incorporated. Although amino acid ethyl esters with β-branched hydrophobic side chains were not effectively incorporated, the more hydrophobic *n*-hexyl esters of these amino acids were more reactive. Similarly, experiments using a peptic hydrolysate of ovalbumin as the substrate indicated a close relationship between the extent of incorporation after 2 hr and the hydrophobicity of the amino acid side chain, except for β-branched amino acids, valine, and isoleucine (Table 8).[161]

The general characteristic of the plastein products is their low solubility in water; gel formation is also frequently observed. In addition, the plastein products often exhibit increased surface hydrophobicity, compared to the protein hydrolysate substrate, as determined by ANS fluorescent probe binding, amount of bound hydrocarbon, binding with the meth-

Table 8
EXTENT OF PAPAIN-CATALYZED
INCORPORATION OF AMINO ACID
ETHYL ESTERS INTO OVALBUMIN
HYDROLYSATE AFTER 2 HR INCUBATION

Amino acid ethyl ester	Velocity of incorporation (μmol/mg papain/min)
Glycine-OEt	0.007
L-Alanine-OEt	0.016
D-Alanine-OEt	0.000
L-α-Aminobutyric acid-OEt	0.058
L-Valine-OEt	0.005
L-Norvaline-OEt	0.122
L-Leucine-OEt	0.119
L-Isoleucine-OEt	0.005
L-Norleucine-OEt	0.125
L-Methionine-OEt	0.115
D-Methionine-OEt	0.000
L-Tyrosine-OEt	0.120
L-Phenylalanine-OEt	0.127
L-Tryptophan-OEt	0.132
L-Glutamic acid-α-OEt	0.025
L-Glutamic acid-α,γ-diOEt	0.109
L-Lysine-OEt	0.043
N^{ϵ}-Acetyl-L-lysine-OEt	0.096

Adapted from Aso et al.[161]

ylene groups of sodium dodecylsulfate determined from nuclear magnetic resonance spectra, and blue shift in tryptophan fluorescence spectra.[162,163]

The plastein reaction has been investigated for many different applications in the food area. It may be used for the removal of unwanted compounds such as odorants.[164-167] Debittering of protein hydrolysates is effective through the plastein reaction.[164,168] The plastein reaction has also been reported for preparation of products of high nutritional quality by incorporating controlled amounts of essential amino acids during the incubation. The plastein reaction was proposed for modification of fish protein concentrate and soybean protein isolate to prepare a peptide-type, low-phenylalanine, high-tyrosine food for patients with phenylketonuria.[169] The nutritional quality of soybean protein was improved by incubating soybean protein hydrolysate with L-methionine ethyl ester and papain under conditions favoring plastein formation; a plastein containing 7.22% methionine, compared to 1.18% methionine in the original protein, was obtained in 70% yield.[170,171] Lysine content was increased by incubation of L-lysine ethyl ester and papain with gluten which had been hydrolyzed with fungal alkaline protease, yielding a plastein with a molecular weight of over 500 and lysine content of 16%.[124]

Although numerous workers have been able to form water-insoluble products through the so-called plastein reaction for resynthesis of peptide bonds, the actual nature of the products and the mechanism for their formation remain controversial. Peptide bond resynthesis may occur either by condensation or by transpeptidation. In the case of condensation, a decrease in free terminal amino groups should occur. Some workers[172] have reported such a decrease, while others[173] have not been able to detect it. At the same time, an increase in the average molecular weight of the peptides should accompany condensation. Some workers have reported an increase in the molecular weights of plastein products compared to their hydrolysates,[174,175] but the validity of these conclusions was considered questionable by van

Hofsten and Lalasidis[176] due to poor solubility of plastein in water and ordinary buffers, making molecular weight determination difficult. Using Sephadex® chromatography[176] and sodium dodecylsulfate electrophoresis,[177] no high molecular weight protein material was detected after plastein reaction.

The other mechanism for peptide bond resynthesis is transpeptidation, which involves an interior peptide bond as a reactant. In this case, there may not necessarily be a decrease in free terminal amino groups, and whether or not there is an increase in the molecular weight depends on the relative sizes of the original peptides involved in the transpeptidation reaction. Horowitz and Haurowitz[173] concluded that transpeptidation was the predominating mechanism in plastein formation by chymotrypsin, yet transpeptidation is hard to accept as the sole factor responsible for the formation of insoluble or gel-forming products.[155]

In lieu of the hypothesis of plastein formation through resynthesis of covalent peptide bonds, some workers have suggested that hydrophobic bonding between peptides is the major force leading to water-insolubility, precipitation or gel formation of plasteins.[155,162,176,177] An increase in the proportion of nonpolar or hydrophobic amino acids in the water-insoluble reaction product in comparison to the original reactant has been reported,[163] and the ratio of hydrophilic to hydrophobic peptides in the incubation mixture for the plastein formation definitely affects the properties of the plasteins.[178] For the plastein reaction to proceed to give good yields, a critical balance between hydrophobic and hydrophilic peptides is required.[170,178] The resulting plasteins are also more hydrophobic than their precursor proteins and they bind apolar compounds more strongly.[9]

In detailed studies on the plastein reaction applied to caseins and skim-milk powder, Sukan and Andrews[179,180] concluded that a physical aggregation mechanism was responsible, principally via hydrophobic bonding but with ionic forces also having a minor role. Hydrophobic amino acids such as phenylalanine, leucine, isoleucine, tyrosine, valine, and proline were preferentially incorporated into the plasteins at the expense of hydrophilic amino acids such as aspartic acid, serine, threonine, glutamic acid, lysine, and arginine. Gel filtration chromatography showed little change in the distribution of peptide molecular weights when plastein was dissolved under dissociating conditions and compared to the peptide sizes in the starting hydrolysate. Peptide-mapping experiments also did not indicate formation of any new peptides or disappearance of existing peptides, casting doubt on the role of condensation or transpeptidation in the plastein reaction. Although the yields (<27%) of the isolated plasteins were too low for economic feasibility of application to the food industry, Sukan and Andrews[180] suggested that exploitation of the plastein reaction could be possible by using the plastein material without separation from accompanying nonincorporated peptides. Formation of products with unique functional properties were anticipated by varying concentrations of peptides and plasteins.

The term "plastein" therefore should be used with caution to describe the enzymatic formation of water-insoluble products from protein or peptide hydrolysate.[176] Like "peptone", the product termed "plastein" may simply be a complex mixture of peptides, the former being a soluble mixture and the latter an insoluble mixture; the insolubility of plastein products may be due to the association of relatively small molecules through noncovalent interactions, particularly hydrophobic interactions. Nevertheless, the possibility of enzyme-catalyzed resynthesis of peptide bonds under suitable conditions should not be discounted. For example, Matsushima et al,[181] recently described the catalysis of peptide-synthesis reaction in benzene using chymotrypsin which had been modified with polyethyleneglycol. These results indicate that by correct matching of hydrophobic parameters of enzyme, substrate, and reaction media, favorable conditions may exist to encourage protease-catalyzed peptide resynthesis reactions.

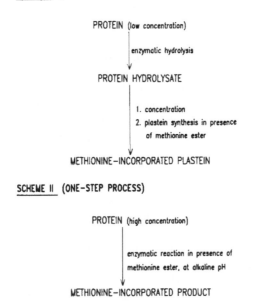

SCHEME I (TWO-STEP:ENZYMATIC DEGRADATION & RESYNTHESIS)

PROTEIN (low concentration)

enzymatic hydrolysis

PROTEIN HYDROLYSATE

1. concentration
2. plastein synthesis in presence
   of methionine ester

METHIONINE-INCORPORATED PLASTEIN

SCHEME II (ONE-STEP PROCESS)

PROTEIN (high concentration)

enzymatic reaction in presence of
methionine ester, at alkaline pH

METHIONINE-INCORPORATED PRODUCT

FIGURE 2.   Comparison of conventional two-step process
and novel one-step process for amino acid incorporation
into food proteins by enzymatic modification.

C. Attachment of Hydrophobic Ligands for Amphiphilicity and Functionality Improvement

In vitro attachment of amino acids to proteins may improve nutritional and functional properties of food proteins. One method of enzymatic incorporation is through transglutaminase-catalyzed substitution of a variety of primary amines including amino acids for the γ-carboxamide groups of protein bound glutaminyl residues. Ikura et al.[182] reported the use of transglutaminase incorporation of methionine ethyl ester to increase methionine content of α_{s1}-casein, β-casein and soybean 7S and 11S proteins to 200, 150, 240, and 350% of the starting material, respectively. In the case of wheat gluten, a 5.1-fold increase in lysine content could be achieved. It was suggested that transglutaminase could be a useful tool for improving the amino acid composition, to improve functional properties and nutritive value. However, drawbacks of this method include the high cost of transglutaminase and the side-reaction of polymerization of protein molecules through formation of ε-(γ-glutamyl) lysine crosslinks.

A novel one-step process was developed as a modified type of plastein reaction by Yamashita et al.[183,184] which allowed papain-catalyzed incorporation of L-methionine and other amino acids directly into soy protein and flour (Figure 2). The mechanism of this process was suggested to be aminolysis of peptide or protein molecules at their carboxyl termini by an amino acid ester acting as the nucleophile. For the enzymatic reaction to proceed efficiently, the following unusual conditions were recommended: (1) use of an alkaline medium at pH 9 to 10, and (2) high concentration of the protein substrate of at least 20% (w/w). Under these specific conditions and controlled ratio of nucleophile (amino acid ester) to substrate (protein), efficient incorporation of the amino acid to the protein by papain-catalyzed aminolysis of protein was the predominant reaction.

Although the novel one-step process for amino acid attachment was first designed with the purpose of nutritional improvement, these Japanese workers soon demonstrated its great potential for improving protein amphiphilicity and functionality as well. It was hypothesized

Table 9
CHANGES IN FUNCTIONALITY OF PROTEINS BY SUCCINYLATION AND LEUCINE ALKYL ESTER ATTACHMENT

Sample[a]	Whippability	Foam stability	Emulsifying activity
Fish protein concentrate			
Original	1.6	0.20	0.00
Succinylated	2.4	0.69	0.57
Leu-OC$_4$	4.4	0.88	0.68
Leu-OC$_{12}$	2.4	0.03	0.70
Soy protein isolate			
Original	1.9	0.15	0.47
Succinylated	2.8	0.36	0.53
Le-OC$_4$	3.9	0.80	0.64
Leu-OC$_{12}$	2.0	0.00	0.64
Casein			
Original	2.5	0.50	0.39
Succinylated	3.4	0.00	0.47
Leu-OC$_4$	4.0	0.55	0.57
Leu-OC$_{12}$	3.2	0.25	0.64
Ovalbumin			
Original	1.2	0.10	0.00
Succinylated	2.4	0.24	0.50
Leu-OC$_4$	3.5	0.80	0.63
Leu-OC$_{12}$	3.4	0.28	0.83
Gelatin			
Original	1.7	0.10	0.00
Succinylated	2.4	0.00	0.23
Leu-OC$_4$	3.4	0.00	0.35
Leu-OC$_{12}$	2.9	0.13	0.53

[a] Leu-OC$_4$ and Leu-OC$_{12}$ were succinylated proteins incubated for 30 min with papain in the presence of leucine alkyl (C$_4$ and C$_{12}$) esters, with the exception of 15 min incubation for succinylated gelatin.

Compiled from data of Watanabe et al.[190]

that by reacting a hydrophilic protein as the substrate with a highly hydrophobic amino acid ester as the nucleophile, a product with amphiphilic properties would result from the localized regions of high hydrophilicity and high hydrophobicity. To obtain adequately hydrophilic proteins as substrates, succinylation could be used to modify the proteins prior to their use as substrates for the one-step process. To obtain adequately hydrophobic nucleophiles, amino acid esters with long chain alkyl groups were used. By this method, these workers successfully produced proteinaceous surfactants with specific functional properties.[127,185-192]

Application of the papain-catalyzed one-step process for L-norleucine n-dodecyl ester attachment to succinylated α_{s1}-casein yielded a surface-active 20,000 dalton product with increased emulsifying activity compared to α_{s1}-casein or succinylated α_{s1}-casein.[185] Subsequently, the process was applied to prepare proteinaceous surfactants with different hydrophile-lipophile balance (HLB) values by using as nucleophiles L-leucine, n-alkyl esters of varying alkyl chain length, and therefore different lipophilicity.[186-191] Commercially available food proteins such as gelatin, fish protein concentrate, soy protein isolate, casein, and ovalbumin were selected as hydrophilic substrates; these proteins were also succinylated to enhance their hydrophilicity.

Succinylation of these proteins generally caused an improvement in whippability, foam stability, and emulsifying activity (Table 9).[190] Further improvement of these functional properties was achieved by incubation of the succinylated proteins with papain in the presence of L-leucine alkyl esters of varying chain length including C_2, C_4, C_6, C_8, C_{10}, and C_{12} (Table 9).[190] For most of the succinylated proteins, the highest whippability resulted when C_4 to C_8 esters of leucine were incorporated, and the highest foam stability resulted on incorporation of C_4 or C_6 esters. When longer chain esters were used, the whippability and foam stability properties were reduced (e.g., Leu-OC_{12} data, Table 9). In contrast, use of increasingly larger alkyl esters of leucine generally resulted in increasing emulsifying activity. An exception was found in the products from succinylated soy protein isolate, whose emulsifying activity was not altered significantly by increasing alkyl length, possibly due to insufficient hydrophilicity of this succinylated protein. Succinylated gelatin also gave products which differed from the general trend. No stable foam was produced by incorporation of C_2 to C_6 esters to succinylated gelatin, and only moderate foam stability resulted by incorporation of the longer alkyl esters. These results were attributed to the much greater hydrophilicity of gelatin itself, compared to the other food proteins under investigation; the highly hydrophilic succinylated gelatin would thus require incorporation of more hydrophobic ligands to improve its amphiphilicity.

Since gelatin was anticipated to be a sufficiently hydrophilic substrate, attempts were carried out to produce proteinaceous surfactants directly from gelatin by the one-step process, without prior succinylation. The enzymatic reaction with papain in the presence of L-leucine alkyl esters produced mixtures of peptides with a wide range of molecular weight, and an average molecular weight of approximately 7500, containing 1.1 to 1.2 mol of covalently attached alkyl moieties per 7500 g product. The surfactancy of the products depended on the alkyl ester incorporated. Incorporation of C_4 to C_6 esters yielded products with good whippability while incorporation of C_{10} to C_{12} esters gave products with a good ability to stabilize oil-in-water emulsions.[186-189,191] These proteinaceous surfactants differed in many respects from conventional surfactants such as Tween-60, Tween-80, and Sunsoft SE-11 (a sucrose fatty acid ester).[186,187] By controlling the proper conditions for emulsification, a variety of functional properties of the emulsions could be observed in terms of hardness, adhesiveness, viscosity, and viscoelasticity. For example, high surfactant concentrations yielded gel formation of the emulsions.

The physicochemical properties of proteinaceous surfactants prepared from gelatin and succinylated fish protein (sFPC) were characterized in terms of hydrophobicity and their ability to interact with water and oil (Table 10).[127] Gelatin-Leu-OC_6 (15 min) and sFPC-Leu-OC_4 (30 min) were selected as whipping surfactants, and gelatin-Leu-OC_{12} (15 min) and sFPC-Leu-OC_4 (30 min) were also selected as whipping surfactants.

Hydrophobicity measured by ANS fluorescence probe increased by attachment of leucine alkyl esters of increasing chain length, suggesting that attachment of an alkyl ester with intermediate degree of hydrophobicity (e.g., C_4 or C_6) could maximize whippability, while a much higher degree of hydrophobicity (e.g., C_{12}) was required for good emulsifying activity.

Electron spin resonance (ESR) measurement of the rotational correlation time (τ_c) of an oil soluble free radical probe of the nitroxide class was used to compare oil binding or immobilization by the surfactants. τ_c increased significantly for both emulsifying and whipping surfactants, compared to their hydrolysate controls, indicating the occurrence of a significant surfactant-oil probe interaction.

Pulsed nuclear magnetic resonance measurement of the spin-spin relaxation times (T_2) of water molecules indicated a significant decrease in T_2 in the dispersions with whipping surfactants (gelatin-Leu-OC_6 and sFPC-Leu-OC_4), compared to the control hydrolysates and the emulsifying surfactants. Decreases in T_2 were observed after whipping and emulsification of dispersions containing whipping and emulsifying surfactants, respectively. Since T_2 can be considered as an index for average mobility of water molecules, with decrease in T_2

Table 10
SOME PHYSICOCHEMICAL PROPERTIES OF PROTEINACEOUS SURFACTANTS PREPARED FROM GELATIN AND SUCCINYLATED FISH PROTEIN CONCENTRATE (FPC)

| | | T_2 (ms) | | | | |
| | | Whipping | | Emulsification | | |
Sample[a]	ANS relative fluorescence intensity	Before	After	Before	After	τ_c ($\times$ 10^{-9} sec)
Gelatin						
Hydrolysate control (15 min)	28	126	—[b]	—	—[c]	1.6
Leu-OC$_6$ (15 min)	60	85	43	81	78	9.3
Leu-OC$_{12}$ (15 min)	381	148	—[b]	145	75	10.3
Succinylated FPC						
Hydrolysate control (15 min)	131	114	—[b]	—	—[c]	1.7
Leu-OC$_6$ (15 min)	146	83	49	74	68	9.4
Leu-OC$_{12}$ (15 min)	696	118	—[b]	140	68	9.9

[a] Samples were incubated for 15 min or 30 min with papain in the absence and presence of leucine alkyl esters, to yield the hydrolysate control and Leu-OC$_6$ or Leu-OC$_{12}$ samples, respectively.
[b] No stable foam was formed.
[c] No stable emulsion was formed.

Adapted from Watanabe and Arai.[127]

indicating increased intensity of binding or immobilization, these results suggest that upon whipping or emulsification, the surfactants were induced to rearrange at air-water or oil-water interfaces, with their hydrophobic alkyl ends oriented towards the air or oil phase and their hydrophilic proteinaceous surfaces oriented towards the aqueous interface, thus restricting mobility of the water molecules. The whipping surfactants appeared to bind water even before whipping, as suggested by their low T_2 values, whereas in the case of the emulsifying surfactants, the large T_2 values before emulsification suggest that these molecules were not effective in preventing water molecules from tumbling until oil was introduced.[127]

Further studies demonstrated the feasibility of the application of these proteinaceous surfactants of enzymatically modified gelatin as substitutes for conventional surfactants in preparing food items. Thus, gelatin with attached leucine C$_2$ to C$_6$ alkyl esters were effective for snow jelly preparation, whereas gelatin with attached leucine C$_{12}$ alkyl ester gave an ice cream product with a high degree of overrun, mayonnaise with a fine emulsion structure, and bread with satisfactory loaf quality and decreased staling upon storage. More recently, emulsions with antifreeze properties were prepared using enzymatically modified gelatin with leucine C$_{12}$ ester.[192] This proteinaceous surfactant could be considered as a bifunctional agent, usable both for production of an oil-in-water emulsion and for cryoprotection against emulsion breakdown upon freezing. These results may indicate further potential application for cryoprotection of cold-sensitive foods and biological systems as well, demonstrating the wide applications arising from enzymatically catalyzed lipophilization of proteins.

IV. ACID OR ALKALI TREATMENT

Acid and alkali can catalyze a number of reactions resulting in changes in the structure and functional properties of proteins. Whitaker[193] listed the following as possible changes occurring in proteins in alkaline solution: denaturation, hydrolysis of some peptide bonds, hydrolysis of amides (asparagine and glutamine), hydrolysis of arginine, some destruction of amino acids, β-elimination and racemization, formation of double bonds, and formation of new amino acids. Proteins are quite susceptible to denaturation in alkaline solution due to decreased stabilization of the tertiary structure by eliminating electrostatic interactions between carboxylate and amino and guanidinium groups, and hydrogen bonding between the hydroxyl group of tyrosine and carboxylate groups.[193] Thus alkaline treatment can accomplish an increased solubilization of the protein, although in some cases upon adjustment of pH back to the neutral or acidic range, the protein may not be soluble due to irreversible denaturation. Acid treatment may also catalyze many of the reactions listed above for alkaline treatment, including hydrolysis of amides and peptide bonds, and destruction of some amino acids and formation of new ones. Frequently, higher solubility is achieved due to higher electrostatic repulsion, decrease in hydrogen bonding, and denaturation. In general, controlled limited treatments with mild acid or alkali are useful in the food industry for improving solubility, emulsifying, and foaming properties.[9] In some cases, extensive hydrolysis may be desirable to produce hydrolysates as flavor enhancers and sources of flavors. However, in these cases, the characteristic functional properties of proteins are sacrificed. More severe conditions may also be used for texturization, fiber spinning, inactivation of enzymes, and destruction of toxins, enzyme inhibitors, and allergens. However, frequently, the more severe conditions, especially if used in conjunction with heating, are undesirable from a nutritional point of view, due to possible formation of indigestible crosslinks or potentially toxic compounds.

Mild acid treatment of food proteins was reported to be effective for improving their functional properties. Since the pioneering work of Holme and Briggs[194] on gliadin, various researchers have reported on the solubilization of wheat gliadin, gluten, and flour by deamidation.[195-199] Finley[196] suggested a mild acid treatment of wheat gluten to increase its solubility in fruit-based acidic beverages. Wu et al.[198] obtained 78% recovery of solubilized vital gluten by heating in dilute (0.02 to 0.05 N) hydrochloric acid or 1.75 N acetic acid, followed by isoelectric precipitation at pH 4.7 to 4.9. Detailed analyses of molecular weight distribution by gel filtration chromatography, approach-to-equilibrium sedimentation analysis, and chemical analyses of amide nitrogen, free amino groups, and sulfhydryl/disulfide groups indicated that solubilization was a result not only of deamidation of glutaminyl and asparaginyl residues but also rupture of a few peptide linkages in gluten molecules. The acid-solubilized gluten had improved emulsifying-capacity and stability properties which were superior to those of soy protein isolate. The fraction extracted at pH 5.2 had excellent foamability and foam-stability properties.

Matsudomi et al.[200] also reported an increase in solubility by acid hydrolysis. However, based on the results of SDS-PAGE, amino acid analysis and gel-filtration chromatography, these workers suggested that solubilization was due mainly to deamidation and not to cleavage of peptide linkages. Emulsifying activity, emulsion stability, and foaming power were improved by deamidation. The improvement in these functional properties was highly correlated with the degree of gluten deamidation, up to 30% deamidation. It was suggested that deamidation may induce a conformational change by increasing electrostatic repulsion and decreasing hydrogen bonding. Subsequently, Matsudomi et al.[201] studied the conformation and surface properties of deamidated gluten. The helix content of gluten decreased curvilinearly with increasing degree of deamidation, whereas the surface hydrophobicity determined using *cis*-parinarate increased in linear proportion to the degree of deamidation, up to 40% deamidation. Surface tension decreased with increasing degree of deamidation and increasing surface hydrophobicity. Emulsifying properties were improved greatly by deamidation and were linearly correlated with surface hydrophobicity. It was concluded that a marked increase in surface hydrophobicity was the main structural factor contributing to improvement in functional properties resulting from deamidation of gluten. The increase in surface hydrophobicity was suggested to induce an amphiphilic nature, decreasing free energy at the surface of the deamidated gluten molecules and thus endowing them with good surface properties.

Changes in conformation and functional properties were also noted for soy protein by mild acid treatment.[202] Controlled acid treatment (0.05 N HCl, 95°C, 30 min) resulted in preferential deamidation without significant cleavage of peptide bonds. Solubility, emulsifying, and foaming properties of the treated soy protein were improved greatly with an increase in surface hydrophobicity at an early stage of the mild acid treatment. It was suggested that the improvement in functionality by controlled acid treatment was due to an increase in surface hydrophobicity caused by deamidation and acid-induced denaturation. Prolonged treatment resulted in poorer functionality, probably due to partial hydrolysis of peptide bonds leading to a decrease in high molecular weight components.

Changes in conformation of soybean proteins in alkaline pH conditions were noted above pH 11.0.[203] Exposure of a previously hydrophobic-buried region was indicated at pH 11.0 by a dramatic increase in binding of a hydrocarbon, n-heptane; however, at pH 12.9, the hydrophobic region was destroyed. Molecular interaction resulting in gel formation during dialysis after alkali treatment above pH 11.0 was suggested to be from intermolecular bonds (hydrogen, hydrophobic, and disulfide) formed by amino acid residues that became accessible by conformational changes during alkali treatment.

Mild acid treatment of ovalbumin resulted in deamidation of asparaginyl and glutaminyl residues as well as hydrolysis of peptide bonds on either side of the aspartate residues.[204] These deamidated ovalbumin peptide fragments obtained by hydrolysis could subsequently be polymerized by standing at room temperature in contact with air. The polymerized products had good functional properties such as solubility, emulsifying activity, emulsion stability, foaming power, and foam stability. These properties were suggested to result from higher electrostatic repulsion due to deamidation, better amphiphilic nature due to proper hydrophobic-hydrophilic arrangement, and polymerization of the amphiphilic fragments through hydrophobic and disulfide bonds.

The driving force for polymerization to form soluble macromolecular aggregates of the ovalbumin peptide fragments was suggested to be hydrophobic and disulfide bond formation.[204] The molecular weight of the polymerized peptides determined by gel filtration on Sephadex® G-100 was decreased by the presence of 0.3 M mercaptoethanol in 0.5% SDS or in 4 M guanidine hydrochloride. Sulfhydryl contents of polymerized ovalbumin peptides decreased to almost one-half compared to the untreated ovalbumin (2.2 to 2.8 mol/mol and 4.0 mol/mol, respectively), while surface hydrophobicity determined using cis-parinarate increased dramatically from 20 for the untreated ovalbumin to between 1720 and 1980 for polymerized samples.[204] It was proposed that the driving force for aggregation in an early stage was hydrophobic, and then after a proper hydrophilic-hydrophobic arrangement had occurred, further polymerization may have occurred by disulfide bond formation, for which closer proximity was a necessity. This polymerization system during mild acid treatment was proposed to be feasible for development of new functional food proteins.

V. MODIFICATION BY PHYSICAL PROCESSES

Physical treatments include the use of thermal energy, mechanical energy, or pressure to modify food proteins. Common examples of physical processes which alter food proteins are heating, freezing, radiation, extrusion, fiber spinning, sonication, whipping to form foams, emulsification with oil, and even extreme dilution. These processes usually result in denaturation of the proteins, i.e., they alter the conformation of the polypeptide chains from the "native" state, without breaking any primary covalent bonds. Denaturation is usually accompanied by unfolding of the protein molecules, without any apparent loss in solubility. The unfolding step is frequently followed by aggregation, which may lead to loss of solubility. Thus, although changes in solubility are frequently used as indicators of protein denaturation, the absence of alteration in solubility should not be interpreted as a lack of denaturation.[205]

Kinsella[9] noted that although the phenomenon of denaturation is familiar, it has different connotations to different researchers. For example, to chemists, biologists, and biochemists, denaturation is undesirable due to alterations which frequently result in loss of solubility and biological activity. However, to food scientists, denaturation is an integral component of various processes for bringing about desirable functional, textural, and sensory properties, for example in the formation of emulsions, foams, gels, and fibers. Thus, while solubility is critical for some applications, controlled denaturation and molecular association arising from physical processes are of practical significance in many other applications. In some cases, controlled physical treatments may improve the property of solubility through partial denaturation and dissociation of molecules.

The chemistry of denaturation of food proteins in particular, and its implications on functional properties, was recently reviewed by Kilara and Sharkasi.[205] According to these authors, denaturation may be considered to be a physical process in the sense that the change from native to other conformations results from physical changes in molecular alignment and intermolecular interactions. These changes require energy input, which may be in the form of heat, light, sound, pressure, or even chemical agents such as urea, guanidine hydrochloride, sodium dodecylsulfate, or salts. For food systems, the most frequently used processes probably involve thermally induced denaturation, and the majority of information on molecular aspects of denaturation of food proteins is thus related to thermal processes. The effects of salts may also be important in some foods. The relative effectiveness of ions in stabilizing proteins against conformational change follows the classical Hofmeister series.[206,207] However, since ionic strength of 0.1 or greater is reported to eliminate or minimize interactions between ions of neutral salts and charged groups of proteins, it may be assumed that under typical conditions found in many foods, electrostatic forces may play a relatively minor role in protein conformation and properties.

A. Thermal Processing for Gelation or Coagulation

The applications of thermal processing for gelation and coagulation of food proteins are well documented.[9,205,208-214] It is generally accepted that the primary techniques for gelation and coagulation in food systems involve heating and/or addition of divalent cations such as calcium.[213] Ferry[215] suggested a two-step mechanism for heat-induced gelation, involving an initiation step of unfolding or dissociation of the protein molecule, followed by an aggregation step in which association and aggregation reactions occurred to form the gel under appropriate conditions. The rate of gel formation appeared to be of great consequence to the hardness of the gels formed.[216] Hermansson[217] noted that a gel network is characterized by a certain degree of order, which may be attained if the second step (association) is slower relative to the first step (unfolding or dissociation), allowing the denatured molecules to rearrange themselves prior to aggregation. If the second step is reversible, as in the case of soy proteins, the intermediate state is defined as a progel state.

Hydrogen bonding, hydrophobic interactions, ionic forces, and disulfide bonds may all be involved in crosslink-bonding of protein gel structures, with particular forces being more dominant in some protein gels than others. For example, hydrogen bonding has been identified as the predominant type of force in gelatin gels,[205] whereas for whey protein concentrates, gelation was reported to be governed by ionic interactions and hydrophobic group availability, with sulfhydryl content playing a relatively minor role.[218] Kornhorst and Mangino[218] reported that for whey protein concentrates, the best model for predicting heat-induced gel strength ($R^2 = 0.93$) was one which included protein hydrophobicity determined by heptane binding and calcium content, with both variables being equally important, whereas sulfhydryl content did not vary sufficiently enough to be a good gel strength predictor. On the other hand, Hillier et al.[219] suggested that whey protein gels were crosslinked mainly by disulfide bonds, with opaque gels being formed from whey powders containing high total sulfhydryl content, and clear gels formed from those with low total sulfhydryl content. However, the rate of gelling was not correlated with total sulfhydryl content. Hydrophobic and electrostatic forces and sulfhydryl-disulfide reactions have been suggested for egg albumen gelation.[39,220] Hayakawa and Nakai[221] suggested that coagulability of ovalbumin was affected by hydrophobicity and net charge with almost no involvement of sulfhydryl groups, whereas gel strength was affected by hydrophobicity and sulfhydryl groups with less involvement of net charge. Utsumi and Kinsella[222] suggested that for soy proteins, electrostatic interactions and disulfide bonds were involved in formation of 11S globulin gels, whereas mostly hydrogen bonding was involved in 7S globulin gels and both hydrogen bonding and hydrophobic interactions were important for gels from soy isolate. The importance of the hydrophobicity of unfolded proteins, rather than surface hydrophobicity, for thermal functional properties of various food proteins, was noted by Voutsinas et al.[223]

Thickening, coagulation, and gelation were significantly correlated with unfolded hydrophobicity and sulfhydryl content.[223] Formation of soluble aggregates of soy protein and existence of sol, progel, gel, and metasol states, and the roles of different forces in their formation have been investigated.[205,224] Beveridge et al.[225,226] used dynamic shear measurements to monitor the kinetics of structural rigidity developed on heating egg white, whey protein concentrate, and soybean protein concentrate. Although the reversibility of progel-gel conversion in soy protein systems has been documented previously, Beveridge et al.[226] have suggested that even for the so-called thermally irreversible gels from protein sources such as egg white and whey, at least partial reversibility can be achieved by reheating of the gels which have been set by cooling. They divided the continuous gelation process for these egg white and whey proteins into three steps following initial denaturation and unfolding of the native protein. In the first step, turbidity is developed, probably directed by hydrophobic interactions and modified by the charged state of the protein molecules. In the second step, sulfhydryl-disulfide reactions stiffen the preformed aggregates and may enhance aggregate adherence. In the third step, hydrogen bonding upon cooling results in a marked increase in elasticity and rigidity.

Despite the ubiquitous presence of coagulated or gelled systems in food products and the abundance of scientific literature examining these phenomena, the molecular basis for these processes are not understood well enough to predict the behavior of different food proteins. As Kinsella[10] noted, the physicochemical basis remains unexplained for the unique properties of various systems such as the excellent thermostable foam-forming properties of egg proteins, the binding and thermosetting functions of myosin in meat systems, and the unique development of cheese texture from the coagulum formed from casein.

B. Thermal Processing for Denaturation without Insolubilization

While heat processes are most commonly used to generate properties resulting from gelation or coagulation, milder heat treatment may also be useful to improve other functional properties such as emulsifying and foaming properties and solubility. For example, whey protein concentrates are endowed with improved whipping properties after partial denaturation with mild heating (e.g., 50°C, 30 min) or by addition of denaturing agents such as sodium dodecylsulfate, prior to the whipping process.[227-233] Mild temperature effects on the structure and solubility of whey proteins in the temperature-range between 4 and 60°C are reversible and governed mainly by the strengthening of hydrophobic bonding and weakening of hydrogen bonding and ionic forces with increasing temperature.[228] It has been suggested that improvement in whipping properties by denaturing treatments is partly due to promotion of protein molecular rearrangements in exposing hydrophobic regions of protein molecules, thus facilitating orientation of the protein at the air/water interface in stabilizing the foam.

Table 11
EFFECTS OF HEATING ON HYDROPHOBICITY, SOLUBILITY, EMULSIFYING, AND FAT-BINDING PROPERTIES OF SOME FOOD PROTEINS

Sample	CPA hydrophobicity	Solubility index, (%)	Emulsifying -activity index, m²/g	Emulsifying -stability index, min	Fat-binding capacity,(%)
Bovine serum albumin					
Control	325	100.0	148	108.5	25.0
100°C, 5 min, pH 4	304	26.8	140	90.0	54.4
β-Lactoglobulin					
Control	426	100.0	96	27.2	4.2
100°C, 15 min, pH 1.0	192	6.4	51	25.3	16.5
Soy protein isolate					
Control	95	26.4	42	6.65	90.0
121°C, 15 min, pH 5.5	160	8.2	50	0.76	68.4
121°C, 15 min, pH 7.2	128	77.2	76	112.5	54.8
Ovalbumin					
Control	6	100.0	24	0.56	42.7
80°C, 3 min, pH 5.6	142	48.9	49	5.85	82.5
100°C, 15 min, pH 1.0	296	46.0	136	12.6	101.3
Canola protein isolate					
Control	65	28.3	60	0.54	33.9
100°C, 2 min, pH 5.5	78	2.9	40	0.14	42.5
121°C, 10 min, pH 7.2	101	21.2	50	2.20	35.7
Whey protein isolate					
Control	182	88.7	87	50.3	74.5
80°C, 4 min, pH 6.0	211	75.2	98	61.3	75.2
80°C, 15 min, pH 6.0	128	50.4	82	48.3	100.8
Casein					
Control	28	100.0	58	1.70	11.3
121°C, 5 min, 0.01 M CaCl₂, pH 7.4	30	76.2	56	9.50	18.1
Gelatin					
Control	5	15.3	46	4.65	19.1
75°C, 2.5 min, pH 7.4	6	100.0	59	10.2	—

Adapted from Voutsinas et al.[234]

Voutsinas et al.[234] studied the effects of heating on changes in hydrophobicity, solubility, emulsifying, and fat-binding properties of some food proteins (Table 11). Quite diverse results were obtained, depending on the protein and the heating conditions. For example, heat treatment of bovine serum albumin greatly reduced its solubility, while hardly affecting hydrophobicity and emulsifying properties. Heating of β-lactoglobulin in highly acidic conditions decreased hydrophobicity, solubility, and emulsifying properties while improving fat-binding capacity. In comparison, intramolecular association without loss of solubility has been reported by heating β-lactoglobulin to 60°C at pH 4.6.[228] Autoclave treatment (121°C) of soy protein isolate at pH 5.5 caused an increase in hydrophobicity, and a large reduction of solubility and emulsion stability, whereas autoclave treatment at pH 7.2 caused a moderate increase in hydrophobicity, and a large increase in solubility and emulsifying properties. A dramatic improvement in solubility of gelatin was observed by heating, and this was accompanied by an improvement in emulsifying properties. Heating ovalbumin solution in highly acidic conditions (pH 1.0) decreased solubility, but greatly increased hydrophobicity, and improved emulsifying properties as well as fat-binding capacity. Unlike soy protein isolate, autoclaving of canola protein isolate at pH 7.2 did not improve solubility or functional properties. For whey protein isolate, a short heat treatment increased hydrophobicity and improved emulsifying properties, but prolonged heating decreased hydrophobicity and solubility, which was accompanied by reduction in emulsifying properties but improvement in fat-binding capacity.

Protein denaturation usually decreases solubility and adversely affects protein functionality. The results in Table 11 demonstrated that for some proteins (e.g., ovalbumin and whey protein isolate) functional properties such as emulsifying and fat-binding properties are improved by heating under specific conditions, despite an accompanying reduction in solubility. For other proteins (e.g., soy protein isolate) solubility can be improved by controlled heat treatment. Voutsinas et al.[234] also showed that emulsifying and fat-binding properties of unheated as well as heat-denatured proteins could be explained and predicted by models incorporating hydrophobicity and solubility variables.

The effects of "partial" denaturation by heating or sodium dodecylsulfate treatment on surface properties of various food proteins, were studied by Kato and co-workers.[235-237] Since most proteins increase surface hydrophobicity as denaturation proceeds, due to exposure to the molecular surface of previously interiorly buried hydrophobic residues, Kato et al.[235] hypothesized that if partial denaturation could be induced so as to increase surface hydrophobicity without formation of a coagulum, the protein functionality, especially surface properties, could be improved. Since denaturation is frequently accompanied by insolubilization, the conditions to induce "partial" denaturation would need to be carefully controlled. Alternatively, these researchers suggested a chemical modification by which the denatured proteins would remain soluble and thus exhibit improved functionality.

The emulsifying and foaming properties of proteins such as ovalbumin and lysozyme were markedly improved by partial denaturation without loss of solubility by specific heating or sodium dodecylsulfate treatments. The increase in surface hydrophobicity of the partially denatured proteins determined by cis-parinaric acid was linearly correlated with the decreasing helix content determined from values of ellipticity at 222 nm measured by circular dichroism.[235] Subsequently, Kato et al.[236] and Matsudomi et al.[237] reported that surface hydrophobicity as well as emulsifying properties of ovalbumin and 7S globulin, increased with heat denaturation, whereas for β-lactoglobulin and bovine serum albumin, the surface hydrophobicity, as well as emulsifying properties, decreased with heat denaturation. Emulsifying properties were significantly linearly correlated with surface hydrophobicity for five proteins subjected to varying extents of heat denaturation, suggesting that the surface hydrophobicity of the proteins was the main factor contributing to their emulsifying properties. On the other hand, foaming power was curvilinearly related to hydrophobicity while no significant correlation was found with foam stability. It was suggested that the improvements in emulsifying properties by denaturation in the absence of insolubilization, may be attributed to the amphiphilic nature of the denatured proteins introduced by the increased surface hydrophobicity. However, in the case of foaming properties, other structural factors in addition to surface hydrophobicity may be important.

Cherry and McWatters[238] reported that ingredients such as peanut, soybean and whey-containing proteins, which had been heat denatured without any accompanying precipitation showed enhanced foamability. They summarized factors affecting foaming properties of proteins, including properties of the proteins such as structural properties (e.g., flexibility, order, availability of hydrophobic and hydrophilic groups, electrostatic forces), denaturation (ease of unfolding), and the Marongonic effect (ability to concentrate rapidly at a stress point in the film); also important were environmental factors such as temperature, pH, viscosity, and presence of other components such as denaturants, which would affect the intrinsic properties of the proteins.

Kato et al.[239] hypothesized that in addition to surface hydrophobicity, the molecular flexibility and susceptibility of proteins to surface denaturation may be involved in functional properties such as emulsifying and foaming properties. Susceptibility of proteins to digestion by proteolytic enzymes was used as an indicator of protein molecular flexibility. Ovalbumin and lysozyme were found to be nonsusceptible to protease digestion, suggesting rigid or folded molecules, whereas κ-casein, β-lactoglobulin and bovine serum albumin were susceptible to protease, suggesting flexible molecules. Good correlations were observed between foaming power and emulsifying activity and digestion velocity of the proteins. Townsend and Nakai[240] also suggested that for quick concentration and subsequent denaturation at a surface or interface, protein molecules should be flexible, structurally less-ordered, and above all, hydrophobic. The molecules would thus be able to undergo conformational changes at the surface by unfolding to expose more hydrophobic regions, facilitating association of the unfolded polypeptide chains and therefore increasing foam stability. However, excessive self-association would be undesirable, due to loss of elasticity of the protein film and decrease in foam stability. Townsend and Nakai[240] reported that foam capacity of various food proteins was not correlated to surface hydrophobicity of the native molecules, but to hydrophobicity measured after treatment at 100°C in the presence of 1.5% sodium dodecylsulfate. Thus, foaming properties could be predicted by the parameter of hydrophobicity of the unfolded or denatured molecules.

C. Sonication

Sonication has been a widely used physical process for solubilizing animal and plant tissue components by virtue of a cavitation effect. It is generally believed that globular proteins may be physically disrupted or transformed under an ultrasonic power field.[241,242] However, its application for improving functionality in food systems and the molecular changes in the sonicated proteins have only been recently investigated. Childs and Forté[243] reported that sonication treatment solubilized over 90% of the proteins from both native and autoclaved cottonseed flour, but only 15 to 17% of the proteins were solubilized from cottonseed meal. In contrast, a combined enzyme-chemical extraction method solubilized over 60% of the protein from cottonseed meal. Combination of sonication treatment with the enzyme-chemical method resulted in an efficient technique for solubilization of protein from screw-expressed cottonseed meal, with at least 70% solubilization at low expenditures of time, enzyme, and energy. It was postulated that ultrasonication may have produced cavitation of the sample, exposing new sites for enzyme action, and also may have increased mixing efficiency, thus improving overall efficiency of the enzyme-chemical treatment.

Wang[244,245] reported that sonication solubilized over 80% of the protein from autoclaved soybean flakes and over 90% of the proteins from alcohol-washed flakes in a single extraction. In comparison, only 16% of the proteins were solubilized from autoclaved flakes using the conventional stir-in-water method. Further studies using gel filtration, electrophoresis and ultracentrifugal sedimentation analyses showed that sonication caused agglomeration or aggregation of soy proteins, particularly the 7S fraction.[246] These aggregates were soluble, as indicated by the improvement in solubilization by sonication. Wang[246] proposed that further studies on the mechanism of aggregate formation by sonication may reveal several possibilities. For example, ultrasonic action may promote hydrophobic interaction of globular proteins in water or may induce formation of complex mixtures as in the case of apolipoproteins, or may alter the equilibrium of protein-protein or protein-lipid interactions to favor the formation of a cluster-type of structure.

The use of sonication for complex formation between phosphatidylcholine and soybean proteins has been reported.[248-250] The behavior of the hydrophobic region was studied using ANS; for both 7S and 11S globulins, the fluorescent intensity of ANS-protein complexes was gradually diminished as the ratio of phosphatidylcholine to protein was increased. Conformational changes were also noted on binding of phosphatidylcholine to the proteins. Results of studies with spin-labeled phosphatidylcholine and sucrose gradient density centrifugation indicated both hydrophobic and hydrophilic interactions. It was suggested that

Table 12
PROPERTIES OF SOLUBLE, SONICATED,
AND REDUCED OVOMUCINS

| | Ovomucin sample | | |
Property	Soluble	Sonicated	Reduced
Hydrophobicity (S_o)	22	48	90
Intrinsic viscosity (mℓ/g)	340	135	45
Foaming properties			
Power ($\mu\tau\zeta$/cm)	1174	465	373
Stability (%)	95	60	54
Emulsifying properties			
Activity (A_{500})	1.09	1.25	1.40
Stability (min)	10	20	38

Adapted from Kato et al.[252]

energy from ultrasonication was required for the association of phosphatidylcholine with soy protein, and that two types of interaction were involved, one based on hydrophobic interaction and one involving binding of phosphatidylcholine lamellae to the protein surface. Kamat et al.[251] noted that soy globulins required prior dissociation with denaturing agents (e.g., urea or guanidine hydrochloride), acid pH or alkaline pH, yielding predominantly a population of subunits with increased surface area and greater proportion of hydrophobic residues in order to activate the globulins to form lipoproteins of characteristic buoyant densities. It is possible that ultrasonication treatment also induces phosphatidylcholine association by exposure of hydrophobic surfaces.

Furukawa and Ohta[241] reported on ultrasonic-induced dispersion of heat-treated, acid-precipitated soy proteins, as intermediates in the commercial production of isolated proteins from defatted soybeans. Changes in flow properties by sonication were suggested from gel filtration studies to be derived from ultrasonic-induced dissociation of the protein aggregates which had been formed by heat treatment. The ultrasonic-induced structural alteration was considered to be associated with a partial cleavage of intermolecular hydrophobic interactions, and not from cleavage of peptide or disulfide bonds. It was postulated that ultrasonically exposed hydrophobic regions were subsequently buried through rearrangement of the molecular structure, conferring a more hydrophilic nature. Salting out of heat-treated acid-precipitated soy protein in a variety of sodium salts corresponded with their molal surface tension and was shown to follow the lyotropic or Hofmeister series, suggesting the predominant effect of the hydrophobic interaction between the proteins and salts and the minor role of electrostatic interaction. Changes in the molecular conformation of acid-precipitated soy protein by thermal processing which resulted in low solubility in salt solution were attributed to exposure of hydrophobic regions. Ultrasonication resulted in a dramatic improvement in solubility of heat-treated acid-precipitated soy protein in 0.5 M NaCl, from 43 to 93% solubility before and after sonication, respectively.

The alterations in structural (hydrophobicity and viscosity) and functional (foaming and emulsifying) properties of ovomucin by sonication and reduction were studied by Kato et al.[252] Ovomucin is the glycoprotein believed to be responsible for the gel-like structure of thick egg white. It is insoluble at neutral pH in nondenaturing media, but may be solubilized by mechanical treatments such as homogenization and sonication under mildly alkaline conditions, or by chemical treatment with denaturing and reducing agents. Soluble ovomucin obtained by homogenization in mildly alkaline buffer has a molecular weight of 8,300,000; the molecular weight is decreased by sonication and reduction to 1,100,000 and 230,000, respectively. Table 12 shows the properties of these three types of ovomucin. Surface hydrophobicity determined with *cis*-parinaric acid increased in the order of soluble < sonicated < reduced ovomucin. Since the dissociation into units with smaller molecular weight also followed this order, it was speculated that hydrophobic interaction may be involved in the association of native ovomucin. Intrinsic viscosity decreased in the order of soluble > sonicated > reduced ovomucin. Good correlations were observed between the foaming properties and intrinsic viscosity, and between the emulsifying properties and surface hydrophobicity.

D. "Salting in/Hydrophobic Out" Process

Murray et al.[253,254] described a novel noncovalent approach to processing and utilizing plant proteins such as fababean protein, by formation of a viscous gelatinous mass called "protein micellar mass". These researchers described a process in which dilution of a high salt protein extract of the fababeans (60 to 80 mg protein per mℓ extract containing 0.3 M sodium chloride) into tap water resulted in a milky white system, which appeared to consist of discrete spherical structures believed to be protein micelles formed by hydrophobic interaction. Upon standing, these micelles coalesced to form the protein micellar mass. Murray et al.[253,254] postulated the mechanism behind the formation of the protein micellar mass to be a "salting in" during extraction with high salt, followed by a "hydrophobic out" phenomenon during the subsequent dilution into tap water. Studies on the effects of pH variations on micelle formation in the isolated vicilin protein from fababeans indicated the most extensive micelle responses were observed from pH 6.0 to 6.8, while a gradual decrease in the degree of protein self-association occurred with increasing pH from 6.8 to 8.0.[255] These results were correlated with significant decreases in vicilin surface hydrophobicity determined by *cis*-parinaric acid, from 296 at pH 6.5 to 158 at pH 8.0. By comparison to the surface hydrophobicity value of 500 upon exposure of vicilin to a higher pH of 10.0, which would correspond to the unfolding of the molecule, the conformational changes from pH 6.0 to 8.0 appeared to reflect an increased burial of hydrophobic residues, which appeared to suppress hydrophobic interactions and also micelle formation. The observed importance of the relative exposure of hydrophobic surface residues was interpreted to be supportive of the premise that these micelle structures were products of hydrophobic associative forces.

E. Fiber Spinning and Thermoplastic Extrusion

Processes such as fiber spinning and thermoplastic extrusion have been applied to modify food systems to yield specific textural and sensory properties.[9] In fiber spinning, an alkaline dispersion of concentrated protein ("dope") is extruded through a wet spinning spinnerette, and the extruded protein filaments are coagulated through an acid-sodium chloride bath; the coagulated fibers are then elongated by stretching and heat treatment. As an alternative to using alkaline dopes, heating of moist proteins may be used to produce denatured protein preparations prior to spinning. The physicochemical basis of fiber formation has been generally accepted as unfolding and uncoiling of peptide chains, followed by parallel realignment and crosslinking. However, there have been few systematic studies on the chemistry and physics of spinning, or on the differences in capability of proteins in fiber formation. The mechanism of texturization by thermoplastic extrusion of proteins has also been postulated as denaturation, followed by realignment and new structure formation.

Although food extrusion has been practiced for nearly 50 years, the technology has been continually developed to the point where the food extruder may now be considered as a high-temperature-short-time bioreactor operating at high temperature with short residence times, under high pressures and high shear forces.[256] The extrusion process thus has many functions, including transport, grinding, hydration, shearing, homogenization, mixing, compression, degassing, thermal treatment, compaction, agglomeration, pumping, orientation of molecules or aggregates, shaping, expansion, formation of porous and/or fibrous textures, and partial drying.[257] Thus, applications of these texturization processes have expanded greatly in the last few decades from the initial roles of mixing and forming, for example in the early application to macaroni production, to present applications for texturization of many defatted vegetable protein ingredients and production of fibrous structures, meat analogs, snack foods, and novelty items.

Although the technology of physical processes such as protein extrusion has been rapidly developing, basic information concerning the chemical and physical changes occurring in components of raw ingredients is still scarce.[258] It is believed that protein unfolding followed by realignment occur, but the mechanisms of these processes are unclear. This basic scientific knowledge would contribute to improving operating efficiency of extruders and serve as a basis for producing texturized protein foods, with desired characteristics for specific product applications from a wide variety of raw ingredients. To date, only limited reports have been published on the possible forces involved in protein insolubilization and structure development during extrusion.

Burgess and Stanley[259] postulated that reaggregation of denatured soy proteins by extrusion into the characteristic texture and microstructure was produced mainly by intermolecular peptide bonds. On the other hand, the insolubility of extruded field bean and soybean proteins were reported by Jeunink and Cheftel[260] to be due mainly to noncovalent interactions and disulfide bonds. Resolubilization of the extruded proteins from soybean[261,262] and from cowpea, mung bean, and soybean[263] by sodium dodecylsulfate and reducing agents such as dithiothreitol or 2-mercaptoethanol have been suggested to support the hypothesis that noncovalent forces (hydrogen, ionic and hydrophobic) and sulfhydryl-disulfide interchange reactions are the major chemical changes occurring during extrusion texturization. Extrusion of soy flour yielded a more regular honeycombed structure than soya isolate, and the soy flour extrudates exhibited a unique phenomenon characterized by minimum hydratability, pH, and carbohydrate solubility at 30 to 40°C, which was suggested to arise from additional stabilization by hydrophobic interactions between the soy proteins and embedded carbohydrates in the extrudates.[261]

Holay and Harper,[262] in studying the influence of extrusion shear environment on texturized soy protein, reported that increasing shear strain and temperature-time in the screw tended to enhance crosslinking between the protein molecules, while increasing shear through the die tended to disrupt the linkages. These results pointed out the need for research to describe the restructuring of protein in shear fields.[257] Cheftel[264] also noted that specific conditions are necessary to obtain satisfactory texturation. While the involvement of new hydrogen, hydrophobic and disulfide bondings in protein insolubilization and texturization is clear, the exact texturation mechanisms are unknown. Cheftel[264] stressed the need for more research on new applications of extrusion for modification of proteins, such as covalent attachment of fatty acids and improvement in functional properties through mild mechanical/thermal treatment, which may be achieved through a food extruder. Clearly more systematic investigations on the changes at the molecular level, including the role of hydrophobic interactions, are required to keep up with and improve the technology of extrusion processing.

VI. EXAMPLE OF QSAR APPROACH AND OPTIMIZATION OF FUNCTIONALITY IMPROVEMENT BY MODIFICATION

A. Preparation of Egg White Substitutes by Whey Protein Modification

The characteristic properties of egg white proteins which make them invaluable in food products such as angel food cake are their excellent foaming properties and ability to form an elastic gel upon heating. In addition, egg white proteins are characterized by their capability to form thermostable foams. Over the years, numerous researchers have reported on the modification of whey proteins with the goal of preparing suitable substitutes for egg

Table 13
COMPARISON OF STRUCTURAL AND FUNCTIONAL PROPERTIES OF WHEY PROTEIN CONCENTRATES (WPC), EGG WHITE, AND WPC MODIFIED BY POLYPHOSPHATE OR PEPSIN TREATMENT

Property	WPC[a]		Egg white powder	Modified WPC[a]	
	1	2		Polyphosphate	Pepsin
Structural					
$H\phi_{ave}$[b]	840		980	—	
CPA hydrophobicity					
Surface, S_o	242.4	193.4	117.3	295.5	180.8
Exposed, S_e	511.8	453.0	1198.4	993.8	458.5
ANS hydrophobicity					
Surface, S_o	64.9	75.3	39.3	78.1	68.7
Exposed, S_e	125.3	118.2	254.8	147.5	101.5
Sulfhydryl content, μM/g protein	21.4	19.7	48.3	28.8	19.7
Disulfide content, μM/g protein	192.0	192.4	77.6	202.8	197.2
Functional					
Overrun, %	93	100	176 (67-83)[c]	60	177
Foam stability, mℓ/5 min	6.5	6.0	3.0 (17.0)[c]	18.0	1.5
Gel strength, N	1.55	1.29	4.15 (2.32)[c]	4.55	0.93

[a]   Two sources of WPC were investigated; WPC-1 was used for polyphosphate treatment, and WPC-2 was used for pepsin treatment.
[b]   Calculated from data for the major proteins in whey (β-lactoglobulin, α-lactalbumin) and egg white (ovalbumin, conalbumin and ovomucoid).
[c]   Numbers in parentheses are the corresponding properties for raw egg white.

Adapted from Nakai and Li-Chan,[265] and To et al.[266]

whites in food formulations.[265,266] For example, in attempts to improve foaming properties, treatments of whey protein involving heat process, cysteine, alcohol, linoleate, proteinases, polyphosphate, activated carbon, and thiolation have been investigated. To improve gelling properties, treatments with pyrophosphate, sodium dodecylsulfate, oxalate, polyphosphate, and heating with acid, alkali, or sulfite have been used.

According to To et al.,[266] the interest in whey proteins as egg white substitutes is understandable because very few other proteins posess the characteristic properties of egg white proteins, such as high sulfhydryl and disulfide contents, heat coagulability, and foaming capability. Table 13 compares some of the structural and functional properties of whey proteins to egg white proteins. The average hydrophobicity values calculated on the basis of amino acid compositions of the major proteins in whey and in egg white are not greatly different between the two protein sources. On the other hand, using both ANS and CPA, surface hydrophobicity of whey proteins was higher than egg white proteins while the hydrophobicity measured after a denaturing treatment (termed "exposed" hydrophobicity or S_e) was higher for egg white than for whey proteins. Sulfhydryl content of whey proteins was lower while disulfide content was higher than the egg white proteins. It was thus anticipated that the foaming and gelling properties of whey protein concentrate could be improved if some of the disulfide groups were reduced and more hydrophobic sites were made accessible, to effectively increase the exposed hydrophobicity values. Thus, a variety of treatments were applied in attempts to improve the functionality of whey protein concentrate, and the structural and functional properties of the selected treatments for foaming improvement and for gelling improvement are shown in Table 13.

Based on preliminary tests, simplex optimization technique was applied using pepsin, cysteine, heating temperature, pH, and incubation time at 35°C as the factors, to optimize the foaming or whipping properties as indicated by percent overrun and foam stability. These experiments indicated that the highest overrun score involved addition of cysteine (1% of protein) and pepsin (2.5% of protein) to a 10% solution of whey protein concentrate, incubating for 52 min at 35°C and pH 2.0, followed by holding at 41°C for 5 min. This treatment gave a product with 177 to 183% overrun, compared to 176% for egg white powder (Table 13, pepsin data). However, foam stability of this pepsin treated whey protein concentrate was slightly reduced, and gel strength was also poor.

Modifications by heat treatment in the presence of alkali or pyrophosphates were successful for improving gel strength of whey protein concentrates. However, pyrophosphate treatment resulted in rough chalky gels and also required high calcium, while alkali heat treatment may be undesirable due to the possibility of forming lysinoalanine or racemization. Polyphosphate treatment was thus chosen for optimization studies to obtain modified whey protein concentrates with maximum gel strength. These experiments showed that a product with gel strength comparable to egg white gels could be produced by heating a 1% whey protein concentrate solution with sodium hexametaphosphate (12 to 15% based on protein) at 55.5°C for 2 min. The gel strength of the polyphosphate-treated whey protein concentrate was 4.55 N, compared to 4.15 for egg white (Table 13). Foam stability was also improved from 6.5 min to 18 min; however, percent overrun was poor.

Comparison of the structural properties of the modified and control whey protein concentrates to egg white (Table 13) suggests that the substantial increase in the exposed hydrophobicity of whey proteins by polyphosphate treatment may be the mechanism behind improvement in gel strength. Polyphosphate may have been able to dissociate protein molecules, allowing exposure of tightly buried hydrophobic sites, which could interact in the formation of a strong gel network. The increase in exposed hydrophobic surfaces may also have enhanced greater intermolecular interaction at the protein-air interface, resulting in the observed enhancement in foam stability. Improvements in emulsifying, fat-binding and water-binding capacities were also reported by polyphosphate treatment of whey protein concentrate.[266] On the other hand, pepsin treatment resulted in slight decreases in hydrophobicity, with little change in sulfhydryl/disulfide contents. It was suggested that the pepsin treatment may have improved foaming capacity measured by percent overrun by an improvement in solubility.

Attempts to maximize both overrun and gel strength simultaneously were unsatisfactory, probably because whipping and gelling properties were incompatible. As shown in Figure 3, when the two modified protein preparations with largest overrun and gel strength properties were mixed in varying proportions, it was found that an increase in the proportion of polyphosphate-treated whey protein increased gel strength while decreasing overrun and foam stability; conversely, increasing the proportion of pepsin-treated whey preparation decreased gel strength while improving the percent overrun. A 75:25 blend of polyphosphate-treated: pepsin-treated whey protein concentrate yielded a product with 87 to 104% overrun and gel strength of 4.2 N, comparable to the properties of raw egg white and egg white powder. However, the modified whey protein concentrate products were unsuitable as complete replacers for egg white in angel food cake formulations, due to depressed cake volume and coarse texture. It was suggested that further work dealing with parameters related to the characteristic thermal stability of egg white foams may offer a more successful approach to understanding the unique functionality of egg white in a food system such as angel food cake. Quantitation of these structural parameters will aid in the search for suitable modification processes for potential egg white replacers such as whey protein.

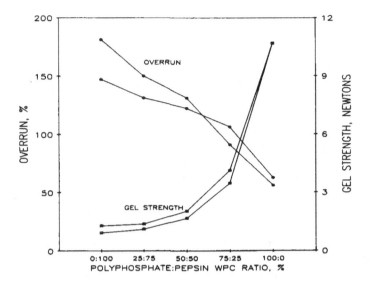

FIGURE 3.   Overrun (○,●) and gel strength (□,■) of modified whey protein concentrates (WPC) formulated wtih varying ratios of polyphosphate-treated WPC to pepsin-treated WPC. Shaded and open symbols represent two different sources of WPC as starting material (Adapted from To et al.[266] *Can. Inst. Food Sci. Technol. J.*, 18, 150, 1985. With permission.)

REFERENCES

1. **Pour-El, A.,** Protein functionality: classification, definition, and methodology, in *Protein Functionality in Foods,* Cherry, J. P., Ed., *ACS Symp. Ser. 147,* American Chemical Society, Washington, D.C., 1981, 1.
2. **Feeney, R. E.,** Chemical changes in food proteins, in *Proteins for Humans: Evaluation and Factors affecting Nutritional Value,* Bodwell, C. E., Ed., AVI, Westport, Conn., 1977, 233.
3. **Feeney, R. E.,** Chemical modification of food proteins, in *Food Proteins. Improvement through Chemical and Enzymatic Modification,* Feeney, R. E. and Whitaker, J. R., Eds., *Adv. Chem. Ser. 160,* American Chemical Society, Washington, D.C., 1977, 1.
4. **Ryan, D. S.,** Determinants of the functional properties of proteins and protein derivatives in foods, in *Food Proteins. Improvement through Chemical and Enzymatic Modification,* Feeney, R. E. and Whitaker, J. R., Eds., *Adv. Chem. Ser. 160,* American Chemical Society, Washington, D.C., 1977, 67.
5. **Cherry, J. P.,** Ed., *Protein Functionality in Foods, ACS Symp. Ser. 147,* American Chemical Society, Washington, D.C., 1981.
6. **Cherry, J. P.,** Ed., *Food Protein Deterioration. Mechanisms and Functionality, ACS Symp. Ser. 206,* American Chemical Society, Washington, D.C., 1982.
7. **Feeney, R. E. and Whitaker, J. R.,** Eds., *Food Proteins, Improvement Through Chemical and Enzymatic Modification, Adv. Chem. Ser. 160,* American Chemical Society, Washington, D.C., 1977.
8. **Feeney, R. E. and Whitaker, J. R.,** Eds., in *Modification of Proteins. Food, Nutritional and Pharmacological Aspects, Adv. Chem. Ser. 198,* American Chemical Society, Washington, D.C., 1982.
9. **Kinsella, J. E.,** Functional properties of proteins: A survey, *Crit. Rev. Food Sci. Nutr.,* 7, 219, 1976.
10. **Kinsella, J. E.,** Relationships between structure and functional properties of food proteins, in *Food Proteins,* Fox, P. F. and Condon, J. J., Eds., Applied Science Publishers, London, 1982, 3.
11. **Pour-El, A.,** Ed., *Functionality and Protein Structure, ACS Symp. Ser. 92,* American Chemical Society, Washington, D.C., 1979.
12. **Hirs, C. H. W. and Timasheff, S. N.,** Modification reactions, *Meth. Enzymol.,* 25B, 385, 1972.
13. **Means, G. and Feeney, R. E.,** Eds., in *Chemical Modification of Proteins,* Holden-Day, San Francisco, 1971.
14. **Singer, S. J.,** Covalent labeling of active sites, *Adv. Protein Chem.,* 22, 1, 1967.

15. **Vallee, B. L. and Riordan, J.**, Chemical approaches to the properties of active sites of enzymes, *Ann. Rev. Biochem.*, 38, 733, 1969.
16. **Feeney, R. E., Yamasaki, R. B., and Geoghegan, K. F.**, Chemical modification of proteins: An overview, in *Modification of Proteins. Food, Nutritional, and Pharmacological Aspects*, Feeney, R. E. and Whitaker, J. R., Eds., *Adv. Chem. Ser. 198*, American Chemical Society, Washington, D.C., 1982, 1.
17. **Kester, J. J. and Richardson, T.**, Modification of whey proteins to improve functionality, *J. Dairy Sci.*, 67, 2757, 1984.
18. **Kinsella, J. E. and Shetty, K. J.**, Chemical modification for improving functional properties of plant and yeast proteins, in *Functionality and Protein Structure*, Pour-El, A., Ed., *ACS Symp. Ser. 92*, American Chemical Society, Washington, D.C., 1979, 3.
19. **Meyer, E. W. and Williams, L. D.**, Chemical modification of soy proteins, in *Food Proteins, Improvement through Chemical and Enzymatic Modification*, Feeney, R. E. and Whitaker, J. R., Eds., *Adv. Chem. Ser. 160*, American Chemical Society, Washington, D.C., 1977, 3.
20. **Nakai, S. and Powrie, W. D.**, Modification of proteins for functional and nutritional improvements, in *Cereals: A Renewable Resource. Theory and Practice*, Pomeranz, Y. and Munck, L., Eds., The American Association of Cereal Chemists, St. Paul, Minn., 1981, 11.
21. **Richardson, T.**, Chemical modifications and genetic engineering of food proteins, *J. Dairy Sci.*, 68, 2753, 1985.
22. **Shukla, T. P.**, Chemical modification of food proteins, in *Food Protein Deterioration. Mechanisms and Functionality*, Cherry, J. P., Eds., *ACS Symp. Ser. 206*, American Chemical Society, Washington, D.C., 1982, 11.
23. **Evans, M., Iron, L., and Petty, J. H. P.**, Physicochemical properties of some acyl derivatives of β-casein, *Biochim. Biophys. Acta*, 243, 259, 1971.
24. **Hoagland, P. D.**, Acylated β-caseins. Effect of alkyl group size on calcium ion sensitivity and on aggregation, *Biochemistry*, 7, 2542, 1968.
25. **Thompson, L. U. and Reyes, E. S.**, Modification of heat-coagulated whey protein concentrates by succinylation, *J. Dairy Sci.*, 63, 715, 1980.
26. **Chen. L., Richardson, T., and Amundson, C.**, Some functional properties of succinylated proteins from fish protein concentrate, *J. Milk Food Technol.*, 38, 89, 1975.
27. **Groninger, H., Jr. and Miller, R.**, Preparation and aeration properties of an enzyme-modified succinylated fish protein, *J. Food Sci.*, 40, 327, 1975.
28. **Miller, R. and Groninger, H. S., Jr.**, Functional properties of enzyme-modified acylated fish protein derivatives, *J. Food Sci.*, 41, 268, 1976.
29. **McElwain, M., Richardson, T., and Amundson, C.**, Some functional properties of succinylated single cell protein concentrate, *J. Milk Food Technol.*, 38, 521, 1975.
30. **Franzen, K. L. and Kinsella, J. E.**, Functional properties of succinylated and acetylated soy protein, *J. Agric. Food Chem.*, 24, 788, 1976.
31. **Canella, M., Castriotta, G., and Bernardi, A.**, Functional and physicochemical properties of succinylated and acetylated sunflower protein, *Lebensm.-Wiss. u.-Technol.*, 12, 95, 1979.
32. **Franzen, K. L. and Kinsella, J. E.**, Functional properties of succinylated and acetylated leaf protein, *J. Agric. Food Chem.*, 24, 914, 1976.
33. **Johnson, E. A. and Brekke, C. J.**, Functional properties of acylated pea protein isolates, *J. Food Sci.*, 48, 722, 1983.
34. **Childs, E. A. and Park, K. K.**, Functional properties of acylated glandless cottonseed flour, *J. Food Sci.*, 41, 713, 1976.
35. **Grant, D. R.**, The modification of wheat flour proteins with succinic anhydride, *Cereal Chem.*, 50, 417, 1973.
36. **Gandhi, S. K., Schultz, J. R., Boughey, F. W., and Forsythe, R. H.**, Chemical modification of egg white with 3,3-dimethylglutaric anhydride, *J. Food Sci.*, 33, 163, 1968.
37. **King, A. J., Ball, H. R., and Garlich, J. D.**, A chemical and biological study of acylated egg white, *J. Food Sci.*, 46, 1107, 1981.
38. **Sato, Y. and Nakamura, R.**, Functional properties of acetylated and succinylated egg white, *Agric. Biol. Chem.*, 41, 2163, 1977.
39. **Ma, C.-Y. and Holme, J.**, Effect of chemical modifications on some physicochemical properties and heat coagulation of egg albumen, *J. Food Sci.*, 47, 1454, 1982.
40. **Ma, C.-Y., Poste, L. M., and Holme, J.**, Effect of chemical modifications on the physicochemical and cake-baking properties of egg white, *Can. Inst. Food Sci. Technol. J.*, 19, 17, 1986.
41. **Montejano, J. G., Hamann, D. D., Ball, H. R., Jr., and Lanier, T. C.**, Thermally induced gelation of native and modified egg white — rheological changes during processing; final strengths and microstructure, *J. Food Sci.*, 49, 1249, 1984.
42. **Shetty, K. J. and Rao, M. S. N.**, Effect of succinylation on the oligomeric structure of arachin, *Int. J. Peptide Protein Res.*, 11, 305, 1978.

43. **Ma, C.-Y.,** Functional properties of acylated oat protein, *J. Food Sci.,* 49, 1128, 1984.
44. **Brekke, C. J. and Eisele, T. A.,** The role of modified proteins in the processing of muscle foods, *Food Technol.,* 35(5), 231, 1981.
45. **Eisele, T. A. and Brekke, C. J.,** Chemical modification and functional properties of acylated beef heart myofibrillar proteins, *J. Food Sci.,* 46, 1095, 1981.
46. **Habeeb, A. F., Cassidy, M. G., and Singer, H.,** Molecular structural effects produced in proteins by reaction with succinic anhydride, *Biochim. Biophys. Acta,* 29, 587, 1958.
47. **Hollecker, M. and Creighton, T. E.,** Effect on protein stability of reversing the charge on amino groups, *Biochim. Biophys. Acta,* 701, 395, 1982.
48. **Gray, C. J. and Lomath, A. W.,** Effect of citraconylation on the thermal aggregation of human serum albumin, *Int. J. Biol. Macromol.,* 2, 2, 1980.
49. **Hoagland, P. D.,** Acylated β-caseins: Effect of alkyl size on calcium ion sensitivity and on aggregation, *Biochemistry,* 7, 2542, 1968.
50. **Hoagland, P. D., Boswell, R. T., and Jones, S. B.,** Preparation and properties of amino-acetylated β-casein, *J. Dairy Sci.,* 54, 1564, 1971.
51. **Evans, M. T. A., Irons, L., and Petty, J. H. P.,** Physicochemical properties of some acyl derivatives of β-casein, *Biochim. Biophys. Acta,* 243, 259, 1971.
52. **Evans, M. T. A., Irons, L., and Jones, M.,** Physicochemical properties of β-casein and some carboxyacyl derivatives, *Biochim. Biophys. Acta,* 229, 411, 1971.
53. **Kim, S. C., Olson, N. F., and Richardson, T.,** The effect of thiolation on β-lactoglobulin, *J. Dairy Sci.,* 66(suppl.1), 98, 1983.
54. **Sung, H.-Y., Chen, H.-J., Liu, T.-Y., and Su, J.-C.,** Improvement of the functionality of soy protein by introduction of new thiol groups through a papain-catalyzed acylation, *J. Food Sci.,* 708, 48, 1983.
55. **Fraenkel-Conrat, H. and Olcott, H. S.,** Esterification of proteins with alcohols of low molecular weight, *J. Biol. Chem.,* 161, 259, 1945.
56. **Mattarella, N. and Richardson, T.,** Adsorption of positively-charged β-lactoglobulin derivatives to casein micelles, *J. Dairy Sci.,* 65, 2253, 1982.
57. **Mattarella, N. L., Creamer, L. K., and Richardson, T.,** Amidation or esterification of bovine β-lactoglobulin to form positively charged proteins, *J. Agric. Food Chem.,* 31, 968, 1983.
58. **Mattarella, N. L. and Richardson, T.,** Physicochemical and functional properties of positively charged derivatives of bovine β-lactoglobulin, *J. Agric. Food Chem.,* 31, 972, 1983.
59. **Lewis, S. D. and Shafer, J. A.,** Conversion of exposed aspartyl and glutamyl residues in proteins to asparaginyl and glutaminyl residues, *Biochim. Biophys. Acta,* 303, 284, 1973.
60. **Hayakawa, S. and Nakai, S.,** Relationships of hydrophobicity and net charge to the solubility of milk and soy proteins, *J. Food Sci.,* 50, 486, 1985.
61. **Ma, C.-Y. and Nakai, S.,** Chemical modification of carboxyl groups in porcine pepsin, *J. Agric. Food Chem.,* 28, 834, 1980.
62. **Ma, C.-Y. and Nakai, S.,** Carboxyl-modified porcine pepsin: properties and reactions on milk and caseins, *J. Dairy Sci.,* 63, 705, 1980.
63. **Means, G. E. and Feeney, R. E.,** Reductive alkylation of amino groups in proteins, *Biochemistry,* 7, 2192, 1968.
64. **Olson, N. F., Richardson, T., and Zadow, J. G.,** Reductive methylation of lysine residues in casein, *J. Dairy Res.,* 45, 69, 1978.
65. **Rowley, B. O., Lund, D. B., and Richardson, T.,** Reductive methylation of β-lactoglobulin, *J. Dairy Sci.,* 62, 533, 1979.
66. **Fretheim, K., Iwai, S., and Feeney, R. E.,** Extensive modification of protein amino groups by reductive addition of different sized substituents, *Int. J. Pept. Protein Res.,* 14, 451, 1979.
67. **Sen, L. C., Lee, H. S., Feeney, R. E., and Whitaker, J. R.,** *In vitro* digestibility and functional properties of chemically modified casein, *J. Agric. Food Chem.,* 29, 348, 1981.
68. **Lee, H. S., Sen, L. C., Clifford, A. J., Whitaker, J. R., and Feeney, R. E.,** Preparation and nutritional properties of caseins covalently modified with sugars. Reductive alkylation of lysines with glucose, fructose, or lactose, *J. Agric. Food Chem.,* 27, 1094, 1979.
69. **Aoki, H., Taneyama, D., Orimo, N., and Kitagawa, I.,** Effect of lipophilization of soy protein on its emulsion stabilizing properties, *J. Food Sci.,* 46, 1192, 1981.
70. **Braun, V.,** Covalent lipoprotein from the outer membrane of *Escherichia coli, Biochim. Biophys. Acta,* 415, 335, 1975.
71. **Henderson, L. E., Krutsch, H. C., and Oroszlan, S.,** Myristyl amino-terminal acylation of murine retrovirus proteins: An unusual post-translational protein modification, *Proc. Natl. Acad. Sci. U.S.A.,* 80, 339, 1983.
72. **Carr, S. A., Biemann, K., Shoji, S., Parmelee, D. C., and Titani, K.,** n-Tetradecanoyl is the NH$_2$-terminal blocking group of the catalytic subunit of cyclic AMP-dependent protein kinase from bovine cardiac muscle, *Proc. Natl. Acad. Sci. U.S.A.,* 79, 6128, 1982.

73. **Torchilin, V. P., Omel'Yanenko, V. G., Klibanov, A. L., Mikhailov, A. I., Gold'Danskii, V. I., and Smirnov, V. N.**, Incorporation of hydrophilic protein modified with hydrophobic agent into liposome membrane, *Biochim. Biophys. Acta*, 602, 511, 1980.

74. **Haque, Z. and Kito, M.**, Lipophilization of soybean glycinin: covalent attachment to long chain fatty acids, *Agric. Biol. Chem.*, 46, 597, 1982.

75. **Haque, Z., Matoba, T., and Kito, M.**, Incorporation of fatty acid into food protein: palmitoyl soybean glycinin, *J. Agric. Food Chem.*, 30, 481, 1982.

76. **Haque, Z. and Kito, M.**, Lipophilization of α_{s1}-casein. 1. Covalent attachment of palmitoyl residue, *J. Agric. Food Chem.*, 31, 1225, 1983.

77. **Haque, Z. and Kito, M.**, Lipophilization of α_{s1}-casein. 2. Conformational and functional effects, *J. Agric. Food Chem.*, 31, 1231, 1983.

78. **Haque, Z. and Kito, M.**, Lipophilization of α_{s1}-casein. 3. Purification and physicochemical properties of novel amphipathic fatty acyl peptides, *J. Agric. Food Chem.*, 32, 1392, 1984.

79. **Haque, Z. and Kito, M.**, One-step lipophilization of proteins using fatty acyl anhydride, *Agric. Biol. Chem.*, 48, 1099, 1984.

80. **Kobrehel, K. and Bushuk, W.**, Studies of glutenin. X. Effect of fatty acids and their sodium salts on solubility in water, *Cereal Chem.*, 54, 833, 1977.

81. **Hegg, P. O. and Lofquist, B.**, The protective effect of small amounts of anionic detergents on the thermal aggregation of crude ovalbumin, *J. Food Sci.*, 39, 1231, 1974.

82. **Boyer, P. D., Lum, F. G., Ballou, G. A., Luck, J. M., and Rice, R. G.**, The combination of fatty acids and related compounds with serum albumin. 1. Stabilization against heat denaturation, *J. Biol. Chem.*, 162, 181, 1946.

83. **Boyer, P. D., Ballou, G. A., and Luek, J. M.**, The combination of fatty acids and related compounds with serum albumin. II. Stabilization against urea and guanidine denaturation, *J. Biol. Chem.*, 162, 199, 1946.

84. **Brandt, J. and Anderson, L.-O.**, Heat denaturation of human serum albumin. Migration of bound fatty acids, *Int. J. Peptide Protein Res.*, 8, 33, 1976.

85. **Nakai, S., Ho, L., Tung, M. A., and Quinn, J. R.**, Solubilization of rapeseed, soy and sunflower protein isolates by surfactant and proteinase treatments, *Can. Inst. Food Sci. Technol. J.*, 13, 14, 1980.

86. **Nakai, S., Ho, L., Helbig, N., Kato, A., and Tung, M. A.**, Relationship between hydrophobicity and emulsifying properties of some plant proteins, *Can. Inst. Food Sci. Technol. J.*, 13, 23, 1980.

87. **Jones, L. J. and Tung, M. A.**, Functional properties of modified oilseed protein concentrates and isolates, *Can. Inst. Food Sci. Technol. J.*, 16, 57, 1983.

88. **Paulson, A. T., Tung, M. A., Garland, M. R., and Nakai, S.**, Functionality of modified plant proteins in model food systems, *Can. Inst. Food Sci. Technol. J.*, 17, 202, 1984.

89. **Kadam, K. L.**, Reverse micelles as a bioseparation tool, *Enzyme Microb. Technol.*, 8, 266, 1986.

90. **Wold, F.**, *In vivo* chemical modification of proteins (Post-translational modification), *Ann. Rev. Biochem.*, 50, 783, 1981.

91. **Kinsella, J. E.**, Functional properties of proteins: possible relationships between structure and function in foams, *Food Chem.*, 7, 273, 1981.

92. **Christensen, T. B., Vegarud, G., and Birkeland, A. J.**, Stabilization of enzymes by glycosylation, *Process. Biochem.*, 11, 25, 1976.

93. **Marshall, J. J. and Rabinowitz, M. L.**, Enzyme stabilization by covalent attachment of carbohydrate, *Arch. Biochem. Biophys.*, 167, 777, 1975.

94. **Kitabatake, N., Cuq, J. L., and Cheftel, J. C.**, Covalent binding of glycosyl residues to β-lactoglobulin: effects on solubility and heat stability, *J. Agric. Food Chem.*, 33, 125, 1985.

95. **Waniska, R. D. and Kinsella, J. E.**, Preparation of maltosyl, β-cyclodextrinyl, glucosaminyl and glucosamineoctaosyl derivatives of β-lactoglobulin, *Int. J. Pept. Protein Res.*, 23, 573, 1984.

96. **Gray, G. R.**, The direct coupling of oligosaccharides to proteins and derivatized gels, *Arch. Biochem. Biophys.*, 163, 426, 1974.

97. **Nakamura, S. and Hayashi, S.**, A role of the carbohydrate moiety of glucose oxidase: kinetic evidence for protection of the enzyme from thermal inactivation in the presence of sodium dodecyl sulfate, *FEBS Lett.*, 41, 327, 1974.

98. **Back, J. F., Oakenfull, D., and Smith, M. B.**, Increased thermal stability of proteins in the presence of sugars and polyols, *Biochemistry*, 18, 5191, 1979.

99. **Waniska, R. D. and Kinsella, J. E.**, Foaming and physicochemical properties of glycosylated derivatives of β-lactoglobulin, Poster 282, 42nd Ann. IFT Meet., Las Vegas, 1982.

100. **Waniska, R. D. and Kinsella, J. E.**, Physicochemical properties of maltosyl and glycosaminyl derivatives of β-lactoglobulin, *Int. J. Pept. Protein Res.*, 23, 467, 1984.

101. **Waniska, R. D. and Kinsella, J. E.**, Surface properties of β-lactoglobulin: Adsorption and rearrangement during film formation, *J. Agric. Food Chem.*, 33, 1143, 1985.

102. **Waniska, R. D. and Kinsella, J. E.**, Enzymatic hydrolysis of maltosyl and glucosaminyl derivatives of β-lactoglobulin, *J. Agric. Food Chem.*, 32, 1042, 1984.

103. **Matheis, G., Penner, M. H., Feeney, R. E., and Whitaker, J. R.**, Phosphorylation of casein and lysozyme by phosphorus oxychloride, *J. Agric. Food Chem.*, 31, 379, 1983.

104. **Sung, H.-Y., Chen, H.-J., Liu, T.-Y., and Su, J.-C.**, Improvement of the functionalities of soy protein isolate through chemical phosphorylation, *J. Food Sci.*, 48, 716, 1983.

105. **Woo, S. L., Creamer, L. K., and Richardson, T.**, Chemical phosphorylation of bovine β-lactoglobulin, *J. Agric. Food Chem.*, 30, 65, 1982.

106. **Woo, S. L. and Richardson, T.**, Functional properties of phosphorylated β-lactoglobulin, *J. Dairy Sci.*, 66, 984, 1983.

107. **Niwa, E., Nakayama, T., and Hamada, I.**, Effect of arylation for setting of muscle proteins, *Agric. Biol. Chem.*, 45, 341, 1981.

108. **Niwa, E., Suzuki, R., Sato, K., Nakayama, T., and Hamada, I.**, Setting of flesh sol induced by ethylsulfonation, *Bull. Jpn. Soc. Sci. Fish.*, 47, 915, 1981.

109. **Niwa, E., Nakayama, T., and Hamada, I.**, Arylsulfonyl chloride induced setting of dolphinfish flesh sol, *Bull. Jpn. Soc. Sci. Fish.*, 47, 179, 1981.

110. **Lindsay, R. C.**, Other desirable constituents of food, in *Principles of Food Science. Part I. Food Chemistry*, Fennema, O. R., Ed., Marcel Dekker, New York, 1976, Chap. 10.

111. **Seguchi, M. and Matsuki, J.**, Studies on pan-cake baking. I. Effect of chlorination of flour on pan-cake qualities, *Cereal Chem.*, 54, 287, 1977.

112. **Seguchi, M.**, Oil-binding capacity of prime starch from chlorinated wheat flour, *Cereal Chem.*, 61, 241, 1984.

113. **Seguchi, M.**, Comparison of oil-binding ability of different chlorinated starches, *Cereal Chem.*, 61, 244, 1984.

114. **Seguchi, M.**, Model experiments on hydrophobicity of chlorinated starch and hydrophobicity of chlorinated surface protein, *Cereal Chem.*, 62, 166, 1985.

115. **Bjarnason-Baumann, B., Pfaender, P., and Silbert, G.**, Enhancement of the biological value of whey protein by covalent addition into peptide linkage of limiting essential amino acids, *Nutr. Metabol.* (Suppl. 1) 21, 170, 1977.

116. **Puigserver, A. J., Sen, L. C., Gonzales-Flores, E., Feeney, R. E., and Whitaker, J. R.**, Covalent attachment of amino acids to casein. 1. Chemical modification and rates of *in vitro* enzymatic hydrolysis of derivatives, *J. Agric. Food Chem.*, 27, 1098, 1979.

117. **Puigserver, A. J., Sen, L. C., Clifford, A. J., Feeney, R. E., and Whitaker, J. R.**, Covalent attachment of amino acids to casein. 2. Bioavailability of methionine and N-acetylmethionine covalently linked to casein, *J. Agric. Food Chem.* 27, 1286, 1979.

118. **Puigserver, A. J., Gaertner, H. F., Sen, L. C., Feeney, R. E., and Whitaker, J. R.**, Covalent attachment of essential amino acids to proteins by chemical methods: Nutritional and functional significance, in *Modification of Proteins. Food, Nutritional and Pharmacological Aspects*, Feeney, R. E. and Whitaker, J. R., Eds., *Adv. Chem. Ser. 198*, American Chemical Society, Washington, D.C., 1982, Chap. 5.

119. **Voutsinas, L. P. and Nakai, S.**, Covalent binding of methionine and tryptophan to soy protein, *J. Food Sci.*, 44, 1205, 1979.

120. **Li-Chan, E., Helbig, N., Holbek, E., Chan, S., and Nakai, S.**, Covalent attachment of lysine to wheat gluten for nutritional improvement, *J. Agric. Food Chem.*, 27, 877, 1979.

121. **Li-Chan, E. and Nakai, S.**, Covalent attachment of N^ϵ-acetyl lysine, N^ϵ-benzylidene lysine and threonine to wheat gluten for nutritional improvement, *J. Food Sci.*, 45, 514, 1980.

122. **Li-Chan, E. and Nakai, S.**, Comparison of browning in wheat glutens enriched by covalent attachment and addition of lysine, *J. Agric. Food Chem.*, 29, 1200, 1981.

123. **Li-Chan, E. and Nakai, S.**, Nutritional evaluation of covalently lysine enriched wheat gluten by *Tetrahymena* bioassay, *J. Food Sci.*, 46, 1840, 1981.

124. **Fujimaki, M., Arai, S.,. and Yamashita, M.**, Enzymatic protein degradation and resynthesis for protein improvement, in *Food Proteins. Improvement through Chemical and Enzymatic Modification*, Feeney, R. E. and Whitaker, J. R., Eds., *Adv. Chem. Ser. 160*, American Chemical Society, Washington, D.C., 1977, Chap. 6.

125. **Phillips, R. D. and Beuchat, L. R.**, Enzyme modification of proteins, in *Protein Functionality in Foods*, Cherry, J. P., Ed., *ACS Symp. Ser. 147*, American Chemical Society, Washington, D.C., 1981, Chap. 13.

126. **Richardson, T.**, Functionality changes in proteins following action of enzymes, in *Food Proteins. Improvement Through Chemical and Enzymatic Modification*, Feeney, R. E. and Whitaker, J. R., Eds., *Adv. Chem. Ser. 160*, American Chemical Society, Washington, D.C., 1977, Chap. 7.

127. **Watanabe, M. and Arai, S.**, Proteinaceous surfactants prepared by covalent attachment of L-leucine n-alkyl esters to food proteins by modification with papain, in *Modification of Proteins. Food, Nutritional and Pharmacological Aspects*, Feeney, R. E. and Whitaker, J. R., Eds., *Adv. Chem. Ser. 198*, American Chemical Society, Washington, D.C., 1982, Chap. 7.

128. **Whitaker, J. R.**, Enzymatic modification of proteins applicable to foods, in *Food Proteins. Improvement through Chemical and Enzymatic Modification*, Feeney, R. E. and Whitaker, J. R., Eds., *Adv. Chem. Ser. 160*, American Chemical Society, Washington, D.C., 1977, Chap. 5.

129. **Whitaker, J. R. and Puigserver, A. J.**, Fundamentals and applications of enzymatic modifications of proteins: An overview, in *Modification of Proteins. Food, Nutritional, and Pharmacological, Aspects*, Feeney, R. E. and Whitaker, J. R., Eds., *Adv. Chem. Ser. 198*, American Chemical Society, Washington, D.C., 1982, Chap. 2.

130. **Adler-Nissen, J.**, Enzymatic hydrolysis of food proteins, *Process Biochem.*, 12, 18, 1977.

131. **Adler-Nissen, J. and Olsen, H. S.**, The influence of peptide chain length on taste and functional properties of enzymatically modified soy protein, in *Functionality and Protein Structure*, Pour-El, A., Ed., ACS Symp. Ser. 92, American Chemical Society, Washington, D.C., 1979, Chap. 7.

132. **Kamata, Y., Ochiai, K., and Yamauchi, J.**, Relationship between the properties of emulsion systems stabilized by soy protein digests and the protein conformation, *Agric. Biol. Chem.*, 48, 1147, 1984.

133. **Zakaria, F. and McFeeters, R. F.**, Improvement of the emulsification properties of soy protein by limited pepsin hydrolysis, *Lebensm.-Wiss.u.-Technol.*, 11, 42, 1978.

134. **Hermansson, A. M., Olsson, D., and Holmberg, B.**, Functional properties of proteins for foods — Modification studies on rapeseed protein concentrates, *Lebensm.-Wiss. u.-Technol.*, 7, 176, 1974.

135. **Bhumiratana, S., Hill, C. G., Jr., and Amundson, C. H.**, Enzymatic solubilization of fish protein concentrate in membrane reactors, *J. Food Sci.*, 42, 1016, 1977.

136. **Hermansson, A. M., Sivik, B., and Skjoldebrand, C.**, Functional properties of proteins for foods — Factors affecting solubility, foaming and swelling of fish protein concentrate, *Lebensm.-Wiss. u.-Technol.*, 4, 201, 1971.

137. **Spinelli, J., Koury, B., and Miller, R.**, Approaches to the utilization of fish for the preparation of protein isolates. Enzymic modification of myofibrillar fish protein, *J. Food Sci.*, 37, 604, 1972.

138. **Dubois, M. W., Anglemier, A. F., Montgomery, M. W., and Davidson, W. D.**, Effect of proteolysis on the emulsification characteristics of bovine skeletal muscle, *J. Food Sci.*, 37, 27, 1972.

139. **Smith, D. M. and Brekke, C. J.**, Functional properties of enzymatically modified beef heart protein, *J. Food Sci.*, 49, 1525, 1984.

140. **Smith, D. M. and Brekke, C. J.**, Enzymatic modification of the structure and functional properties of mechanically deboned fowl proteins, *J. Agric. Food Chem.*, 33, 631, 1985.

141. **Kuehler, C. A. and Stine, C. M.**, Effect of enzymatic hydrolysis on some functional properties of whey protein, *J. Food Sci.*, 39, 379, 1974.

142. **Draudt, H. N., Whistler, R. L., Babel, F. J., and Reitz, H.**, Modification of wheat protein for preparation of milk-like products, *J. Agric. Food Chem.*, 13, 407, 1965.

143. **Yang, H.-J. and McCalla, A. G.**, Action of proteolytic enzymes on wheat gluten, *Can. J. Biochem.*, 46, 1019, 1968.

144. **Grunden, L. P., Vahedra, D. V., and Baker, R. C.**, Effects of proteolytic enzymes on the functionality of chicken egg albumin, *J. Food Sci.*, 39, 841, 1974.

145. **Puski, G.**, Modification of functional properties of soy proteins by proteolytic enzyme treatment, *Cereal Chem.*, 52, 655, 1975.

146. **Pour-El, A. and Swenson, T. S.**, Gelation parameters of enzymatically modified soy protein isolates, *Cereal Chem.*, 53, 438, 1976.

147. **Fuke, Y., Sekiguchi, M., and Matsuoka, H.**, Nature of stem bromelain treatments on the aggregation and gelation of soybean proteins, *J. Food Sci.*, 50, 1283, 1985.

148. **Ney, K. H.**, Voraussage der Bitterkeit von Peptiden aus deren Aminosäurezusammensetzung. *Z. Lebensm. Unters.-Forsch.*, 147, 64, 1971.

149. **Ney, K. H.**, Aminosäurezusammensetzung von Proteinen und die Bitterkeit ihrer Peptide, *Z. Lebensm. Unters.-Forsch.*, 149, 321, 1972.

150. **Lalasidis, G. and Sjoberg, L.-B.**, Two new methods of debittering protein hydrolysates and a fraction of hydrolysates with exceptionally high content of essential amino acids, *J. Agric. Food Chem.*, 26, 742, 1978.

151. **Schalinatus, E. and Behrke, U.**, Untersuchungen über bittere Peptide aus Casein und Käse. 3. Charakterisierung bitterer Peptide und Untersuchungen zur Ausschaltung des Bittergeschmacks, *Nahrung*, 19, 447, 1975.

152. **Helbig, N. B., Ho, L., Christy, G. E., and Nakai, S.**, Debittering of skimmilk hydrolysate by adsorption for incorporation into acidic beverages, *J. Food Sci.*, 45, 331, 1980.

153. **Sawjalow, W. W.**, Zur Theorie der Eiweissverdauung, *Pflügers Archiv für die gesamte Physiologie*, 85, 171, 1901.

154. **Borsook, H.,** Peptide bond formation, *Adv. Protein Chem.,* 8, 127, 1953.

155. **Eriksen, S. and Fagerson, I. S.,** The plastein reaction and its applications: A review, *J. Food Sci.,* 41, 490, 1976.

156. **Wasteneys, H. and Borsook, H.,** The enzymatic synthesis of protein, *Physiol. Rev.,* 10, 110, 1930.

157. **Determann, H., Bonhard, K., Koehler, R., and Wieland, T.,** Untersuchgen über die Plastein-reaktion, *Helv. Chim. Acta,* 46, 2498, 1963.

158. **Virtanen, A. I.,** Über die enzymatische Polypeptidsynthese, *Makromol. Chem.,* 6, 94, 1951.

159. **Yamashita, M., Tsai, S.-J., Arai, S., Kato, H., and Fujimaki, M.,** Enzymatic modification of proteins in foodstuff. V. Plastein yields and their pH dependence, *Agric. Biol. Chem.,* 35, 86, 1971.

160. **Aso, K., Yamashita, M., Arai, S., and Fujimaki, M.,** *Abstracts of papers,* Annual meeting of the Agricultural Chemical Society of Japan, Sapporo, Japan, 1975, 153, cited by Fujimaki, M., Arai, S. and Yamashita, M., in *Food Proteins. Improvement through Chemical and Enzymatic Modification,* Feeney, R. E. and Whitaker, J. R., Eds., *Adv. Chem. Ser. 160,* American Chemical Society, Washington, D.C., 1977, Chap. 6.

161. **Aso, K., Yamashita, M., Arai, S., and Fujimaki, M.,** *Abstracts of papers,* 47th Meeting of the Biochemical Society of Japan, Okayama, Japan, 1974, 435, cited by Fujimaki, M., Arai, S. and Yamashita, M., in *Food Proteins. Improvement through Chemical and Enzymatic Modification,* Feeney, R. E. and Whitaker, J. R., Eds., *Adv. Chem. Ser. 160,* American Chemical Society, Washington, D.C., 1977, Chap. 6.

162. **Aso, K., Yamashita, M., Arai, S. and Fujimaki, M.,** General properties of a plastein synthesized from a soybean protein hydrolysate, *Agric. Biol. Chem.,* 37, 2505, 1973.

163. **Aso, K., Yamashita, M., Arai, S., Fujimaki, M.,** Hydrophobic force as a main factor contributing to plastein chain assembly, *J. Biochem. (Tokyo),* 76, 341, 1974.

164. **Arai, S., Noguchi, M., Kurosawa, S., Kato, H., and Fujimaki, M.,** Applying proteolytic enzymes on soybean. 6. Deodorization effect of *Aspergillopeptidase* A and debittering effect of *Aspergillus* acid carboxypeptidase, *J. Food Sci.,* 35, 392, 1970.

165. **Arai, S., Noguchi, M., Yamashita, M., Kato, H., and Fujimaki, M.,** Studies on flavor components in soybean. VI. Some evidence for occurrence of protein-flavor binding, *Agric. Biol. Chem.,* 34, 1569, 1970.

166. **Fujimaki, M., Kato, H., Arai, S., and Tamaki, E.,** Applying proteolytic enzymes on soybean. 1. Proteolytic enzyme treatment of soybean protein and its effect on flavor. *Food Technol.,* 22, 889, 1968.

167. **Fujimaki, M., Yamashita, M., Arai, S., and Kato, H.,** Enzymatic modification of proteins in foodstuffs. Part I. Enzymatic proteolysis and plastein synthesis application for preparing bland protein-like substances, *Agric. Biol. Chem.,* 34, 1325, 1970.

168. **Fujimaki, M., Yamashita, M., Okazawa, Y., and Arai, S.,** Plastein reaction. Its application to debittering of proteolyzates, *Agric. Biol. Chem.,* 34, 483, 1970.

169. **Yamashita, M., Arai, S., and Fujimaki, M.,** Plastein reaction for food protein improvement, *J. Agric. Food Chem.,* 24, 1100, 1976.

170. **Arai, S., Yamashita, M., and Fujimaki, M.,** Plastein reaction and its applications, *Cereal Foods World,* 20, 107, 1975.

171. **Yamashita, M., Arai, S., Tsai, S.-J., and Fujimaki, M.,** Plastein reaction as a method for enhancing the sulfur-containing amino-acid level of soybean protein, *J. Agric. Food Chem.,* 19, 1151, 1971.

172. **Wasteneys, H. and Borsook, H.,** The synthesizing action of pepsin, *J. Biol. Chem.,* 62, 15, 1924.

173. **Horowitz, J. and Haurowitz, F.,** Mechanism of plastein formation, *Biochim. Biophys. Acta,* 33, 231, 1959.

174. **Yamashita, M., Arai, S., Kokubo, S., Aso, K., and Fujimaki, M.,** A plastein with an extremely high amount of glutamic acid, *Agric. Biol. Chem.,* 38, 1269, 1974.

175. **Tauber, H.,** Synthesis of protein-like substances by chymotrypsin, *J. Am. Chem. Soc.,* 73, 1288, 1951.

176. **van Hofsten, B. and Lalasidis, G.,** Protease-catalyzed formation of plastein products and some of their properties, *J. Agric. Food Chem.,* 24, 460, 1976.

177. **Edwards, J. H. and Shipe, W. F.,** Characterization of plastein reaction products formed by pepsin, α-chymotrypsin, and papain treatment of egg albumin hydrolysates, *J. Food Sci.,* 43, 1215, 1978.

178. **Arai, S., Yamashita, M., Fujimaki, M., and Aso, K.,** A parameter related to the plastein formation, *J. Food Sci.,* 40, 342, 1975.

179. **Sukan, G. and Andrews, A. T.,** Application of the plastein reaction to caseins and to skim-milk powder. I. Protein hydrolysis and plastein formation, *J. Dairy Res.,* 49, 265, 1982.

180. **Sukan, G. and Andrews, A. T.,** Application of the plastein reaction to caseins and to skim-milk powder. II. Chemical and physical properties of the plasteins and the mechanism of plastein formation, *J. Dairy Res.,* 49, 279, 1982.

181. **Matsushima, A., Okada, M., and Inada, Y.,** Chymotrypsin modified with polyethylene glycol catalyzes peptide synthesis reaction in benzene, *FEBS Lett.,* 178, 275, 1984.

182. **Ikura, K., Yoshikawa, M., Sasaki, R., and Chiba, H.,** Incorporation of amino acids into food proteins by transglutaminase, *Agric. Biol. Chem.,* 45, 2587, 1981.

183. **Yamashita, M., Arai, S., Imaizumi, Y., and Fujimaki, M.,** A one-step process for incorporation of L-methionine into soy protein by treatment with papain, *J. Agric. Food Chem.*, 27, 52, 1979.
184. **Yamashita, M., Arai, S., Ahrano, Y., and Fujimaki, M.,** A novel one-step process for enzymatic incorporation of amino acids into proteins: application to soy protein and flour for enhancing their methionine levels, *Agric. Biol. Chem.*, 43, 1065, 1979.
185. **Arai, S. and Watanabe, M.,** Modification of succinylated α_{s1}-casein with papain: covalent attachment of L-norleucine dodecyl ester and its consequence, *Agric. Biol. Chem.*, 44, 1979, 1980.
186. **Arai, S., Watanabe, M., and Fujii, N.,** Physicochemical properties of enzymatically modified gelatin as a proteinaceous surfactant, *Agric. Biol. Chem.*, 48, 1801, 1984.
187. **Shimada, A., Yazawa, E., and Arai, S.,** Preparation of proteinaceous surfactants by enzymatic modification and evaluation of their functional properties in a concentrated emulsion system, *Agric. Biol. Chem.*, 46, 173, 1982.
188. **Watanabe, M., Toyokawa, H., Shimada, A., and Arai, S.,** Proteinaceous surfactants produced from gelatin by enzymatic modification: Evaluation for their functionality, *J. Food Sci.*, 46, 1467, 1981.
189. **Watanabe, M., Shimada, A., Yazawa, E., Kato, T., and Arai, S.,** Proteinaceous surfactants produced from gelatin by enzymatic modification: Application to preparation of food items, *J. Food Sci.*, 46, 1738, 1981.
190. **Watanabe, M., Shimada, A., and Arai, S.,** Enzymatic modification of protein functionality: Implantation of protein amphiphilicity to succinylated proteins by covalent attachment of leucine alkyl esters, *Agric. Biol. Chem.*, 45, 1621, 1981.
191. **Watanabe, M., Fujii, N., and Arai, S.,** Characterization of foam- and emulsion-stabilizing functions of enzymatically modified proteins with surfactancy, *Agric. Biol. Chem.*, 46, 1587, 1982.
192. **Watanabe, M., Tsuji, R. F., Hirao, N., and Arai, S.,** Antifreeze emulsions produced from linoleic acid and water using enzymatically modified gelatin, *Agric. Biol. Chem.*, 49, 3291, 1985.
193. **Whitaker, J. R.,** Changes occurring in proteins in alkaline solution, in *Chemical Deterioration of Proteins*, Whitaker, J. R. and Fujimaki, M., Eds., *ACS Symp. Ser. 123*, American Chemical Society, Washington, D.C., 1980, 7.
194. **Holme, J. and Briggs, D. R.,** Studies on the physical nature of gliadin, *Cereal Chem.*, 36, 321, 1959.
195. **Aranyi, C. and Hawryleciwz, E. J.,** Preparation and isolation of acid-catalyzed hydrolysates from wheat gluten, *J. Agric. Food Chem.*, 20, 670, 1972.
196. **Finley, J. W.,** Deamidated gluten: a potential fortifier for fruit juices, *J. Food Sci.*, 40, 1283, 1975.
197. **McDonald, C. E. and Pence, J. W.,** Wheat gliadin in foams for food products, *Food Technol.*, 15, 141, 1961.
198. **Wu, C. H., Nakai, S., and Powrie, W. D.,** Preparation and properties of acid-solubilized gluten, *J. Agric. Food Chem.*, 24, 504, 1976.
199. **Fung, C. P., Fong, A. S., Lam, A. S. M., Lee, G., and Nakai, S.,** Preparation and properties of acid-solubilized wheat flour, *J. Food Sci.*, 42, 1594, 1977.
200. **Matsudomi, N., Kaneko, S., Kato, A., and Kobayashi, K.,** Functional properties of deamidated gluten, *Nippon Nogeikagaku Kaishi*, 55, 983, 1981.
201. **Matsudomi, N., Kato, A., and Kobayashi, K.,** Conformation and surface properties of deamidated gluten, *Agric. Biol. Chem.*, 46, 1583, 1982.
202. **Matsudomi, N., Sasaki, T., Kato, A., and Kobayashi, K.,** Conformational changes and functional properties of acid-modified soy protein, *Agric. Biol. Chem.*, 49, 1251, 1985.
203. **Ishino, K. and Okamoto, S.,** Molecular interaction in alkali denatured soybean proteins, *Cereal Chem.*, 52, 9, 1975.
204. **Matsudomi, N., Sasaki, T., Tanaka, A., Kobayashi, K., and Kato, A.,** Polymerization of deamidated peptide fragments obtained with the mild acid hydrolysis of ovalbumin, *J. Agric. Food Chem.*, 33, 738, 1985.
205. **Kilara, A. and Sharkasi, T. Y.,** Effects of temperature on food proteins and its implications on functional properties, *CRC Crit. Rev. Food Sci. Nutr.*, 23, 323, 1986.
206. **von Hippel, P. H. and Schleich, T.,** The effects of neutral salts on the structure and conformational stability of macromolecules in solution, in *Structure and Stability of Macromolecules in Solution*, Vol. 2, Timasheff, S. N. and Fasman, G. D., Eds., Marcel Dekker, New York, 1969, 417.
207. **Schulz, G. E. and Schrimer, R. H.,** *Principles of Protein Structure*, Springer-Verlag, New York, 1979.
208. **Shimada, K. and Matsushita, S.,** Effects of salts and denaturants on thermocoagulation of proteins, *J. Agric. Food Chem.*, 29, 15, 1981.
209. **Shimada, K. and Matsushita, S.,** Relationship between thermocoagulation of proteins and amino acid compositions, *J. Agric. Food Chem.*, 28, 413, 1980.
210. **Busk, G. C., Jr.,** Polymer-water interactions in gelation, *Food Technol.*, 38(5), 59, 1984.
211. **Gossett, P. W., Rizvi, S. S. H., and Baker, R. C.,** Quantitative analysis of gelation in egg protein systems, *Food Technol.*, 38(5), 67, 1984.

212. **Ziegler, G. R. and Acton, J. C.,** Mechanisms of gel formation by proteins of muscle tissue, *Food Technol.,* 38(5), 77, 1984.
213. **Schmidt, R. H.,** Gelation and coagulation, in *Protein Functionality in Foods,* Cherry, J. P., Ed., ACS Symp. Ser. 147, American Chemical Society, Washington, D.C., 1981, Chap. 7.
214. **Schmidt, R. H. and Morris, H. A.,** Gelation properties of milk proteins, *Food Technol.,* 38(5), 85, 1984.
215. **Ferry, J. D.,** Protein gels, *Adv. Protein Chem.,* 4, 1, 1948.
216. **Egelandsdal, B.,** A comparison between ovalbumin gels formed by heat and by guanidinium hydrochloride denaturation, *J. Food Sci.,* 49, 1099, 1984.
217. **Hermansson, A.-M.,** Aggregation and denaturation involved in gel formation, in *Functionality and Protein Structure,* Pour-El, A., Ed., ACS Symp. Ser. 92, American Chemical Society, Washington, D.C., 1979, Chap. 5.
218. **Kornhorst, A. L. and Mangino, M. E.,** Prediction of the strength of whey protein gels based on composition, *J. Food Sci.,* 50, 1403, 1985.
219. **Hillier, R. M., Lyster, R. L. J., and Cheeseman, G. C.,** Gelation of reconstituted whey powders by heat, *J. Sci. Food Agric.,* 31, 1152, 1980.
220. **Shimada, K. and Matsushita, S.,** Thermal coagulation of egg albumin, *J. Agric. Food Chem.,* 28, 409, 1980.
221. **Hayakawa, S. and Nakai, S.,** Contribution of hydrophobicity, net charge and sulfhydryl groups to thermal properties of ovalbumin, *Can. Inst. Food Sci. Technol. J.,* 18, 290, 1985.
222. **Utsumi, S., and Kinsella, J. E.,** Forces involved in soy protein gelation: effects of various reagents on the formation, hardness and solubility of heat-induced gels from 7S, 11S and soy isolate, *J. Food Sci.,* 50, 1278, 1985.
223. **Voutsinas, L. P., Nakai, S., and Harwalkar, V. R.,** Relationships between protein hydrophobicity and thermal functional properties of food proteins, *Can. Inst. Food Sci. Technol. J.,* 16, 185, 1983.
224. **Catsimpoolas, N. and Meyer, E. W.,** Gelation phenomena of soybean globulins. I. Protein-protein interactions, *Cereal Chem.,* 47, 559, 1970.
225. **Beveridge, T., Arntfield, S. D., and Murray, E. D.,** Protein structure development in relation to denaturation temperatures, *Can. Inst. Food Sci. Technol. J.,* 18, 189, 1985.
226. **Beveridge, T., Jones, L., and Tung, M. A.,** Progel and gel formation and reversibility of gelation of whey, soybean and albumen protein gels, *J. Agric. Food Chem.,* 32, 307, 1984.
227. **DeWit, J. N.,** Foaming properties of whey protein concentrate, *Neth. Milk Dairy J.,* 30, 75, 1976.
228. **DeWit, J. N. and Klarenbeek, G.,** Effects of various heat treatments on structure and solubility of whey proteins, *J. Dairy Sci.,* 67, 2701, 1984.
229. **Haggett, T. O. R.,** The whipping, foaming and gelling properties of whey protein concentrates, *New Zealand J. Dairy Sci. Technol.,* 11, 275, 1976.
230. **Morr, C. V.,** Functionality of whey protein products, *New Zealand J. Dairy Sci. Technol.,* 14, 185, 1979.
231. **Morr, C. V.,** Functionality of heated milk proteins in dairy and related foods, *J. Dairy Sci.,* 68, 2773, 1985.
232. **Morr, C. V.,** Composition, physicochemical and functional properties of reference whey protein concentrates, *J. Food Sci.,* 50, 1406, 1985.
233. **Schmidt, R. H., Packard, V. H., and Morris, H. A.,** Effect of processing on whey protein functionality, *J. Dairy Sci.,* 67, 2723, 1984.
234. **Voutsinas, L. P., Cheung, E., and Nakai, S.,** Relationships of hydrophobicity to emulsifying properties of heat denatured proteins, *J. Food Sci.,* 48, 26, 1983.
235. **Kato, A., Tsutsui, N., Matsudomi, N., Kobayashi, K., and Nakai, S.,** Effects of partial denaturation on surface properties of ovalbumin and lysozyme, *Agric. Biol. Chem.,* 45, 2755, 1981.
236. **Kato, A., Osako, Y., Matsudomi, N., and Kobayashi, K.,** Changes in the emulsifying and foaming properties of proteins during heat denaturation, *Agric. Biol. Chem.,* 47, 33, 1983.
237. **Matsudomi, N., Mori, H., Kato, A., and Kobayashi, K.,** Emulsifying and foaming properties of heat-denatured soybean 11S globulins in relation to their surface hydrophobicity, *Agric. Biol. Chem.,* 49, 915, 1985.
238. **Cherry, J. P. and McWatters, K. H.,** Whippability and aeration, in *Protein Functionality in Foods,* Cherry, J. P., Ed., *ACS Symp. Ser. 147,* American Chemical Society, Washington, D.C., 1981, Chap. 8.
239. **Kato, A., Komatsu, K., Fujimoto, K., and Kobayashi, K.,** Relationship between surface functional properties and flexibility of proteins detected by the protease susceptibility, *J. Agric. Food Chem.,* 33, 931, 1985.
240. **Townsend, A.-A. and Nakai, S.,** Relationships between hydrophobicity and foaming characteristics of food proteins, *J. Food Sci.,* 48, 588, 1983.
241. **Furukawa, T. and Ohta, S.,** Ultrasonic-induced modification of flow properties of soy protein dispersion, *Agric. Biol. Chem.,* 47, 745, 1983.
242. **Furukawa, T. and Ohta, S.,** Solubility of isolated soy protein in ionic environments and an approach to improve its profile, *Agric. Biol. Chem.,* 47, 751, 1983.

243. **Childs, E. A. and Forté, J. F.**, Enzymatic and ultrasonic techniques for solubilization of protein from heat-treated cottonseed products, *J. Food Sci.*, 41, 652, 1976.
244. **Wang, L. C.**, Ultrasonic extraction of proteins from autoclaved soybean flakes, *J. Food Sci.*, 40, 549, 1975.
245. **Wang, L. C.**, Ultrasonic peptization of soybean proteins from autoclaved flakes, alcohol-washed flakes and commercial samples, *J. Food Sci.*, 43, 1311, 1978.
246. **Wang, L. C.**, Soybean protein agglomeration: Promotion by ultrasonic treatment, *J. Agric. Food Chem.*, 29, 177, 1981.
247. **Ohtsuru, M., Kito, M., Takeuchi, Y., and Ohnishi, S.**, Association of phosphatidylcholine with soybean protein, *Agric. Biol. Chem.*, 40, 2261, 1976.
248. **Kanamoto, R., Ohtsuru, M., and Kito, M.**, Diversity of soybean protein-phosphatidylcholine complex, *Agric. Biol. Chem.*, 41, 2021, 1977.
249. **Ohtsuru, M., Yamashita, Y., Kanamoto, R., and Kito, M.**, Association of phosphatidylcholine with soybean 7S globulin and its effect on the protein conformation, *Agric. Biol. Chem.*, 43, 765, 1979.
250. **Ohtsuru, M. and Kito, M.**, Association of phosphatidylcholine with soybean 11S globulin, *Agric. Biol. Chem.*, 47, 1907, 1983.
251. **Kamat, V. B., Graham, G. E., and Davis, M. A. F.**, Vegetable protein: lipid interactions, *Cereal Chem.*, 55, 295, 1978.
252. **Kato, A., Oda, S., Yamanaka, Y., Matsudomi, N., and Kobayashi, K.**, Functional and structural properties of ovomucin, *Agric. Biol. Chem.*, 49, 3501, 1984.
253. **Murray, E. D., Myers, C. D., and Barker, L. D.**, Protein product and process for preparing same, *Canadian Patent*, 1,028,552, 1978.
254. **Murray, E. D., Myers, C. D., Barker, L. D., and Maurice, T. J.**, Functional attributes of proteins — A noncovalent approach to processing and utilizing proteins, in *Utilization of Protein Resources*, Stanley, D. W., Murray, E. D., and Lees, D. H., Eds., Food and Nutrition Press, Westport, Conn., 1981, 158.
255. **Ismond, M. A. H., Murray, E. D., and Arntfield, S. D.**, The role of non-covalent forces in micelle formation by vicilin from *Vicia faba*. The effect of pH variations in protein interactions, *Food Chem.*, 20, 305, 1986.
256. **Cheftel, J. C.**, Extrusion cooking. introduction and conclusion, in *Thermal Processing and Quality of Foods*, Zeuthen, P., Cheftel, J. C., Eriksson, C., Jul, M., Leniger, H., Linko, P., Varela, G., and Vos, G., Eds., Elsevier, New York, 1984, 23.
257. **Harper, J. M.**, Recent applications and research perspectives in the field of extrusion cooking, in *Thermal Processing and Quality of Foods*, Zeuthen, P., Cheftel, J. C., Eriksson, C., Jul, M., Leniger, H., Linko, P., Varela, G., and Vos, G., Eds., Elsevier, New York, 1984, 25.
258. **Rhee, K. C., Kuo, C. K., and Lusas, E. W.**, Texturization, in *Protein Functionality in Foods*, Cherry, J. P., Ed., *ACS Symp. Ser. 147*, American Chemical Society, Washington, D.C., 1981, Chap. 4.
259. **Burgess, L. D. and Stanley, D. W.**, A possible mechanism for thermal texturization of soybean protein, *Can. Inst. Food Sci. Technol. J.*, 9, 228, 1976.
260. **Jeunink, J. and Cheftel, J. C.**, Chemical and physicochemical changes in field bean and soybean proteins texturized by extrusion, *J. Food Sci.*, 44, 1322, 1979.
261. **Sheard, P. R., Ledward, D. A., and Mitchell, J. R.**, Role of carbohydrates in soya extrusion, *J. Food Technol.*, 19, 475, 1984.
262. **Holay, S. H. and Harper, J. M.**, Influence of the extrusion shear environment on plant protein texturization, *J. Food Sci.*, 47, 1869, 1982.
263. **Pham, C. B. and de Rosario, R. R.**, Studies on the development of texturized vegetable products by the extrusion process. I. Effect of processing variables on protein properties, *J. Food Technol.*, 19, 535, 1984.
264. **Cheftel, J. C.**, Nutritional effects of extrusion-cooking, *Food Chemistry*, 20, 263, 1986.
265. **Nakai, S. and Li-Chan, E.**, Structure modification and functionality of whey proteins: Quantitative structure-activity relationship approach, *J. Dairy Sci.*, 68, 2763, 1985.
266. **To, B., Helbig, N. B., Nakai, S., and Ma, C. Y.**, Modification of whey protein concentrate to simulate whippability and gelation of egg white, *Can. Inst. Food Sci. Technol. J.*, 18, 150, 1985.

Appendix

PROCEDURES FOR MEASURING PROTEIN HYDROPHOBICITY

TABLE OF CONTENTS

I. FLUORESCENCE PROBE METHODS[1,2]

Hydrophobicity is determined using hydrophobic fluorescence probes, 1-anilino-8-na-phthalenesulfonate (ANS) or *cis*-parinarate (CPA). Magnesium salt of ANS was prepared from the technical grade sodium salt (Eastman Kodak Company, Rochester, New York) according to the method of Weber and Young[3]. CPA was purchased from Molecular Probes (Junction City, Oregon). The purity of CPA is critical in obtaining reproducible results; the crystals should be completely soluble in ethanol. Measurements are performed essentially according to the method of Kato and Nakai[1] in the absence of SDS. Each protein sample (2 mℓ) is serially diluted with 0.01 M phosphate buffer (pH 5.5 to 7.4, containing 0 to 0.6 M NaCl; for most routine tests, pH 7.0 without adding NaCl has been used) to obtain protein concentrations typically ranging from 0.005 to 0.030%. Then, 10 $\mu\ell$ ANS (8.0 mM in 0.1 M phosphate buffer, pH 7) or CPA (3.6 mM in absolute ethanol containing equimolar butylated hydroxytoluene) solution is added. Relative fluorescence intensity (RFI) is measured with a spectrofluorometer at wavelength (λ_{ex}, λ_{em}) of (390, 470) nm and (325, 420)nm for ANS and CPA, respectively. The RFI reading is standarized by adjusting the reading of the fluorometer to 30% full scale for ANS in methanol, and 75% full scale for CPA in decane. The net RFI at each protein concentration is measured by subtracting RFI of each solution without probe from that with probe. The initial slope (S_o) of the net RFI vs. protein concentration (%) plot is calculated by linear regression analysis and used as an index of the protein hydrophobicity.

II. HYDROPHOBIC-INTERACTION AND REVERSED-PHASE CHROMATOGRAPHY[4]

The HPLC instrumentation consisted of a Spectra-Physics SP8700 solvent delivery system and an SP8750 organizer module in conjunction with a Hewlett-Packard HP1040A detection system, HP 3390A integrator, HP85 computer, HP9121 disc drive, and HP47A plotter.

Proteins and peptide analogues were separated on a Bio-Gel TSK-Phenyl-5-PW column (BioRad Laboratories, Richmond, California; 75 × 7.5 mm I.D.). Additional experiments were performed on an Altex Ultrapore RPSC C-3 reversed-phase column (Beckman, Palo Alto, California, 75 × 4.6 mm I.D.) and a TSK G2000SW size-exclusion column (Toyo Soda, Tokyo; 6000 × 7.5 mm I.D.). For RPC, samples were dissolved in 0.1% aqueous Trifluoroacetic acid (TFA) at pH 2.1, and the column was eluted with a gradient constructed from 0.1% aqueous TFA (solvent A) and 0.05% TFA in acetonitrile (solvent B). The gradient program was as follows: linear gradient from 100% A to 49% A, 51% B at 17 min (3% B/min), which was then maintained at this percentage of B for an additional 5 min. For HIC, samples were dissolved in solvent A containing 1.7 M ammonium sulfate and 0.1 M sodium phosphate buffer (38 mM NaH$_2$PO$_4$ and 62 mM Na$_2$HPO$_4$) at pH 7.0; column elution was effected with a linear gradient from 100% A to 100% B containing 0.1 M sodium phosphate buffer (38 mM NaH$_2$PO$_4$ and 62 mM Na$_2$HPO$_4$) at pH 7.0 for 15 min, followed by an additional 5 min at 100% B.

III. HYDROCARBON-BINDING CAPACITY[5]

A. Equilibration of Protein Solution and Hydrocarbon

The buffer used to prepare 2.0% (w/v) protein solutions was 0.1 M sodium phosphate buffer (pH 6.8). Four to eight 10 mm × 45 mm test tubes were each filled with 1.20 mℓ of 2.0% protein solution and 0.90 mℓ of *n*-heptane. The tubes were sealed with corks covered with plastic (Saran®) wrap, and the contents of each were mixed by sideways rotation at 2 rpm in a water bath at 20 ± 0.2°. After equilibration for at least 8 hr, the tubes were removed from the bath, dried quickly, and opened.

After equilibrium was reached, a Pasteur pipette was used to remove the supernatant layer from each tube. Then a volumetric pipette was used to transfer approximately 1 mℓ of the lower phase into a 1 mℓ- conical centrifuge tube which had been previously cooled in an ice-water bath. 1 to 5 $\mu\ell$ of this equilibrated protein solution from each of four or more replicates were analyzed directly by gas-liquid chromatography, as described below. Controls, containing no protein, were also equilibrated and analyzed.

B. Analysis of Bound Hydrocarbon

Since aqueous solutions were injected directly into the gas-liquid chromatographic apparatus (Model A-600D Varian Aerograph), a hydrogen-flame ionization detector was used. The gas-chromatographic column (stainless steel, 6 ft $\times$ 0.125 inch outside diameter) was packed with a diatomaceous support (Chromosorb W, acid-washed, 60 to 80 mesh, F. and M. Scientific Corporation) coated with 10% (w/w) silicone (SF-96, General Electric Company). The packed column was heated overnight at 150° before use. According to the volatility of the organic compound to be analyzed, the column temperature and the injection port temperature were selected in the 80 to 140°, and 230 to 350° ranges, respectively. The H2 flow rate was 20 mℓ/min and the N_2 flow rate 6 mℓ/min. The range selector was set at 1, 10, or 100. Chromatograms were recorded with a 1 mV- full-scale recorder.

The syringe used for injections was a 1- or 5-$\mu\ell$ Hamilton syringe, 7000 series, "plunger-in-needle" type. The usual type of syringe was not satisfactory because the proteins coagulated inside the needle and clogged it.

Calibration curves were made by diluting with CS_2-known amounts of n-heptane, and injecting different quantities of each solution into the gas chromatograph. A plot of peak area vs. quantity of organic compound injected, gave a straight-line correlation for n-heptane.

IV. SDS-BINDING CAPACITY[6]

The SDS-binding capacity was determined as follows: SDS was added in 10 mℓ of a 0.1% protein solution and adjusted to 0.07 mM. After being allowed to stand for 30 min, SDS-protein mixtures were dialyzed against 25 volumes of 0.02 M phosphate buffer, pH 6.0, for 24 hr. The SDS contents of inner dialyzates were determined by Epton's method.[7] Ten milliliters of $CHCl_3$ was added to 0.5 mℓ of inner dialyzates and mixed in a test tube. Then 2.5 mℓ of a 0.0024% methylene blue solution was added to the $CHCl_3$ layer. After being mixed in the test tube, the mixture was centrifuged at 2500 rpm. The absorbance of the SDS-methylene blue mixture in the lower layer was measured at 655 nm. Thus, SDS-binding capacity was determined and was represented as micrograms of SDS bound to 500 μg of protein.

V. TRIGLYCERIDE-BINDING CAPACITY BY GLC[8]

A. Materials

Triglycerides used were: tributyrin, trihexanoin, trioctanoin, trinonanoin, tripalmitolein, triolein, trilinolein, and trilinolenin (Sigma Chemical Company, London, 99%).

Polycarbonate membrane filters (25 mm diameter) and filter holders were purchased from Nuclepore Corporation, Pleasanton, California. Monojet disposable syringes were obtained from Sherwood Medical Industries, Deland, Florida.

B. Method

Emulsion preparation — Each triglyceride suspension was obtained by emulsifying 0.25 g triglyceride in 50 mℓ distilled water with an Ultra Turrax TP 18/2 blender (Ika-Werk, Staufen, West Germany) for 30 sec at a setting of 5. The suspension was sonicated with a

Sonifier Cell Disrupter A180G equipped with a 1.27-cm diameter tip (Ultrasonics, Ltd., PBI, Italy). The sonifier was preheated by operating for 5 min in 50 mℓ water. The settings of the power and tune switches were at 8 and 2, respectively. Each suspension was sonicated for four periods of 30-sec duration separated by 30-sec intervals. The resulting microemulsion was passed through a No. 4 Whatman filter, discarding the last few milliliters along with the filter that retained nonemulsified triglyceride.

Equilibration and filtration of protein-triglyceride mixture — The buffer used to prepare 0.1% (w/v) BSA solution was 0.1 M sodium phosphate, pH 6.8. One volume of emulsion was mixed gently with six volumes of protein-buffer solution (sample) or with six volumes of buffer (blank). The blank was used to check the solubility of different triglycerides in buffer alone. Following a 60-min quiescent equilibration at 20 to 22°C, 14 mℓ of the sample mixture or the blank was drawn into a 15 mℓ syringe. The filter holder with the "O" ring on the syringe side of the 0.4 μm pore size polycarbonate membrane filter was attached to the syringe. The first 3 mℓ of filtrate was discarded and the next 10 mℓ were collected for subsequent extraction. A 10- mℓ aliquot of unfiltered sample mixture was set aside as a control.

Extraction of triglycerides — Sample, control, and buffer blank were heated with a 15 mℓ mixture of ethanol-N HCl (15:1, v/v) in 50- mℓ capped tubes for 5 min at 60°C. Fifteen milliliters of hexane were added and the tubes were then shaken for 30 min in a mechanical shaker. Following phase separation, 10 mℓ of upper phase were removed and dried under nitrogen. Each residue was dissolved in 0.5 mℓ of hexane containing 1 mg/mℓ of tributyrin as internal standard (IS) for GLC.

Evaluation of triglyceride-protein interaction — Triglyceride-protein interaction or binding is considered to be related to the amount of triglyceride present before (control) and after (sample) passage of 10 g of protein-triglyceride mixture through the 0.4 μm Nuclepore filter. Percent binding is defined as the percentage of triglyceride in the sample compared to the unfiltered control. The amounts of triglyceride were determined by GLC. The instrument used was a Varian Aerograph 3700 (Varian Associates, Palo Alto, California) equipped with a flame ionization detector, a 67- cm glass column 3.2-mm I.D. packed with 3% OV1 on Gas Chrom 100/120 mesh (Supelco, Incorporated, Bellefonte, Pennsylvania). Conditions were as follows: on-column injection at 350°C; helium-flow rate, 100 mℓ/min; detector temperature, 350°C. For low molecular weight triglycerides, column temperature was 120°C programmed to 250°C at 10° C/min. High molecular weight triglycerides (tripalmitolein and higher) were determined isothermally at 340°C. Four or more injections of approximately 2 μℓ were made and the results were averaged for each analysis. To obtain quantitative data for each triglyceride, a series of standard solutions was prepared with different concentrations of that triglyceride and a known amount of tributyrin as internal standard.

VI. TRIGLYCERIDE-BINDING CAPACITY BY FLUOROMETRY[9]

From 1 mℓ of DPH-corn oil solution in heptane (10 μg 1, 6-diphenyl-1,3,5-hexatriene from Sigma Chemicals, St. Louis, Missouri and 10 mg oil/mℓ heptane) taken in a 22 × 93-mm flat-bottom test tube, heptane was evaporated by nitrogen flushing. After adding 10 mℓ protein solution (0.02 to 0.12%), the content of the test tube was blended using an Ultra-Turrax (Janke and Kunkel, Staufen, West Germany) at 12,800 rpm for 5 min. while cooling in ice water. After centrifuging the emulsion at 27,000 × g for 30 min and diluting the bottom layer solution 20 times with 0.02 M phosphate buffer pH 7.4, fluorescence intensity (FI) of the solution was measured with an Aminco Bowman spectrophotofluorometer No. 4-8202 at 366 and 450 nm for excitation and emission, respectively. The relative fluorescence intensity (RFI) readings were standardized by adjusting the reading of the fluorometer to

70% full-scale for DPH in heptane. For myoglobin, 0.1 M phosphate buffer, pH 7.4 was used. Fish extract and beef extract were diluted with 0.01 M sodium phosphate buffer pH 6.5, containing 0.6 M NaCl, 1 mM MgCl$_2$ and 0.04% sodium azide. After centrifuging the emulsion, the protein sometimes precipitated. When this happened, about 4.5 mℓ of the bottom layer solution was withdrawn; the precipitate was transferred to another tube; and the bottom layer solution separated as above was added to the precipitate up to the 4.0 mℓ mark, then blended with the Ultra Turrax at 12,800 rpm for 1 min. An aliquot of 0.4 mℓ homogenate was mixed with 0.6 mℓ of bottom layer solution then diluted 20 times with 0.02 M phosphate buffer pH 7.4.

VII. HYDROPHOBIC PARTITION[10]

A. Materials

Dextran T500 $\overline{M}_w$ = 460 · 10^3, and dextran T70, $\overline{M}_w$ = 70 · 10^3 were obtained from Pharmacia Fine Chemicals, Sweden. Poly(ethylene glycol), Carbowax 6000, $\overline{M}_n$ = 6 · 10^3 was obtained from Union Carbide. Poly(ethylene glycol) palmitate was synthesized as described by Shanbhag and Johansson[11] and approximately 60% of the total hydroxyl groups in poly(ethylene glycol) were substituted.

B. Two-Phase System

A system with the following total composition: 8% dextran T70, 8% poly(ethylene glycol), (including ligand derivative), 100 mM K$_2$SO$_4$ and 2 mM potassium phosphate buffer pH 7.1 was used in this work unless otherwise stated. The phase systems were prepared by mixing stock solutions of 20% dextran and 40% poly(ethylene glycol) together with salt solutions and protein solutions at room temperature. After mixing by gentle inversions the systems were centrifuged at 1500 × g for 5 min to speed up the separation. The composition (in %) of the two phases obtained was the following: upper phase: 13% poly(ethylene glycol), 0.3% dextran, 86.7% water; lower phase: 0.9% poly(ethylene glycol), 22% dextran, 77.1% water. The partition coefficient of salts is very close to unity.

C. Partition Coefficient of a Protein

The concentration of the partitioned protein in each phase was determined photometrically. An aliquot was withdrawn from each phase and diluted with water. The absorbance of the diluted phase was measured using as a reference an identically diluted solution of the corresponding phase from a system containing no protein. The absorbance was measured at 280 nm for all proteins.

A measure of hydrophobicity, Δ log K was computed from:

$$\Delta\log K = \log [C_u/C_l]_I - \log [C_u/C_l]_{II}$$

where C_u and C_l are concentrations of the partitioned protein in the upper and lower phases, respectively, for a pair of two-phase systems, I and II.

REFERENCES

1. **Kato, A. and Nakai, S.,** Hydrophobicity determined by a fluorescence probe method and its correlation with surface properties of proteins, *Biochim. Biophys. Acta,* 624, 13, 1980.
2. **Hayakawa, S. and Nakai, S.,** Relationships of hydrophobicity and net charge to the solubility of milk and soy proteins, *J. Food Sci.,* 50, 486, 1985.

3. **Weber, G. and Young, L. B.**, Fragmentation of bovine serum albumin by pepsin. I. The origin of the acid expansion of the albumin molecule, *J. Biol. Chem.*, 239, 1415, 1964.

4. **Ingraham, R. H., Lau, S. Y. M., Taneja, A. K., and Hodges, R. S.**, Denaturation and the effects of temperature on hydrophobic-interaction and reversed-phase high-performance liquid chromatography of proteins. Bio-gel TSK-phenyl-5-PW column, *J. Chromatog.*, 327, 77, 1985.

5. **Mohammadzadeh-K, A., Feeney, R. E., and Smith, L. M.**, Hydrophobic binding of hydrocarbons by proteins. I. Relationship of hydrocarbon structure, *Biochim. Biophys. Acta*, 194, 246, 1969.

6. **Kato, A., Matsuda, T., Matsudomi, N., and Kobayashi, K.**, Determination of protein hydrophobicity using a sodium dodecylsulfate-binding method, *J. Agric. Food Chem.*, 32, 284, 1984.

7. **Epton, S. R.**, A new method for the rapid titrimetric analysis of sodium alkyl sulphates and related compounds, *Faraday Soc., Trans.*, 44, 226, 1948.

8. **Smith, L. M., Fantozzi, P., and Creveling, R. K.**, Study of triglyceride-protein interaction using a microemulsion-filtration method, *J. Amer. Oil Chem. Soc.*, 60, 960, 1983.

9. **Tsuitsui, T., Li-Chan, E., and Nakai, S.**, A simple fluorometric method for fat-binding capacity as an index of hydrophobicity of protein, *J. Food Sci.*, 51, 1268, 1986.

10. **Shanbhag, V. P. and Axelsson, C.-G.**, Hydrophobic interaction determined by partition in aqueous two-phase systems. Partition of protein in systems containing fatty-acid esters of poly(ethylene glycol), *Eur. J. Biochem.*, 60, 17, 1975.

11. **Shanbhag, V. P. and Johansson, G.**, Specific extraction of human serum albumin by partition in aqueous biphasic systems containing poly(ethylene glycol) bound ligand, *Biochem. Biophys. Res. Commun.*, 61, 1141, 1974.

INDEX

M

N

O

Printed and bound by CPI Group (UK) Ltd, Croydon, CR0 4YY

22/10/2024

01777600-0003